Do You Want Cookies?

Trust dynamics and educational gaps in the datafied risk society

Angelica M. Maineri

Open Press Tilburg University

ISBN 978-94-036-7630-2
DOI 10.26116/ET28-VC69

Published by Open Press Tilburg University,
Tilburg, the Netherlands
https://www.openpresstiu.org

Typesetting & design by Thomas F. K. Jorna
JOTE Publishers

Do you want cookies? Trust dynamics and educational gaps in the datafied risk society

Proefschrift ter verkrijging van de graad van doctor
aan Tilburg University
op gezag van de rector magnificus, prof. dr. W.B.H.J. van de Donk, in het openbaar te verdedigen ten overstaan van een door het college voor promoties aangewezen commissie in de Aula van de Universiteit op vrijdag 9 december 2022 om 10.00 uur
door

Angelica Maria Maineri
geboren te Milaan, Italië

Promotor:
prof. dr. P.H.J. Achterberg, Tilburg University

Copromotor:
dr. A.R.C.M. Luijkx, Tilburg University & University of Trento

Promotiecommissie:
prof. dr. G.A. Veltri, University of Trento
prof. dr. Y.J. Park, Howard University
prof. dr. M.E.H. van Reisen, Tilburg University
prof. dr. W. de Koster, Erasmus University Rotterdam
prof. dr. S.M.E. Wyatt, Maastricht University

Abstract

Ulrich Beck described the shift to the Risk Society, characterized by the uneven distribution of manufactured risks stemming from human activity. In the thesis I apply Beck's Risk Society perspective to the study of datafication, with a twofold aim: on the one hand, the Risk Society theory can explain the uneven acknowledgment of risks unfolding within the datafied society; on the other hand, the process of datafication constitutes an interesting case to test the empirical grounding of the reflexive modernization thesis, according to which, as modernization progresses, technological progress is increasingly questioned. Results of four empirical studies show how some elements of the reflexive modernization theory do not pass the empirical test. First, a country's level of digitalization does not deepen knowledge-based stratification mechanisms. Second, individuals' trust in data institutions does not drop when the pitfalls of datafication become visible, challenging the 'worked-and-won' dynamic of trust in the risk society. Nevertheless, I also show how the Risk Society perspective is beneficial to better understand some aspects of the datafication processes, as findings indicate the success of organized irresponsibility dynamics, as well as the important role of knowledge as a risk stratification mechanism at the individual level.

Contents

Abstract ix

List of Figures xv

List of Tables xix

1 Introduction 1

Datafication and Risks 1
The Risk Society 5
The contribution of the thesis: understanding the datafied risk society 7

2 In Zuck We Trust? 23

Introduction 24
Theoretical background 27
Data and methods 32
Results 40
Conclusion and Discussion 50

54

3 Public acceptance of a COVID-19 Health Pass 55

Introduction 56
Accepting the COVID-19 Health Pass 58
The present study 62
Data and Methods 67
Results 72
Conclusion and Discussion 79

4 The Closing Educational Gap in E-privacy Management in European Perspective 85

Introduction 86
The educational gap in e-privacy management 89
Data and methods 92
Results 101
Summary of findings and Discussion 106

5 Switch on the Big Brother! 115

Introduction 116
Explaining acceptance of surveillance 118
Data and Method 125
Results 132
Discussion 136

6 Conclusions 145

The Datafied Risk Society 145

Limitations and future work 154
Concluding remarks 156

A Appendix to In Zuck We Trust? 159

B Appendix to Public acceptance of a COVID-19 Health Pass 167

Test of proportional odds 168
Question wording 169
Additional results 174

C Appendix to The Closing Educational Gap in E-privacy Management in European Perspective 179

Non-Internet users 179
Alternative operationalization of Reflexive mindset 181

D Appendix to Switch on the Big Brother! 185

Bibliography 214

Summary 215

Acknowledgements 219

List of Figures

1.1 Share of internet users by country between 2015 and 2017. Own elaboration of data from Ourworldindata.org. 16

2.1 Timeline of the study (source of Google search volumes: Google Trends). 35

2.2 Correlation plot among institutional trust items. 37

2.3 Predicted probabilities of change in trust in social media and 95% confidence intervals by institutional trust (estimated from the model in Table 2.4). 48

3.1 Predicted Probabilities (with 95% Confidence Intervals) of Willingness to use COVID-19 HP by institutional trust (PPO model), N = 1,464. 73

3.2 Predicted Probabilities of Willingness to use HP with 95% Confidence intervals by experimental conditions, estimated via ordinal logistic model (PPO model), N = 1,454. 76

3.3 Predicted Probabilities of Willingness to use Health Pass and 95% Confidence intervals by experimental conditions by the purpose of use of the HP, estimated via ordinal logistic model (PPO model), N = 1,454. Full models available in Appendix B 77

3.4 Predicted Probabilities of Willingness to use Health Pass and 95% Confidence intervals by experimental conditions by levels of trust in government, estimated via ordinal logistic model (PPO model), N = 1454. Only extreme categories are displayed for parsimony, full results can be found in Appendix B 78
3.5 Predicted Probabilities of Willingness to use Health Pass and 95% Confidence intervals by experimental conditions by levels of trust in science, estimated via ordinal logistic model (PPO model), N = 1454. Only extreme categories are displayed for parsimony, full results can be found in Appendix B 80

4.1 Percentage of mentions for each e-privacy management action. . 94
4.2 Mean of years spent in education by country (error bars represent 95% confidence intervals). 96
4.3 Distribution of Digital Economy and Society Index (DESI) 2017 across EU countries. 100
4.4 Average number of e-privacy management activities by country. The dotted line represents the grand-mean across countries. . . 102
4.5 Predicted values of e-privacy management and 95% confidence intervals by 10th, 50th and 90th percentile of years spent in full-time education and age categories (controlling for all variables included in M5). 105
4.6 Predicted values of e-privacy management and 95% confidence intervals by 10th, 50th and 90th percentile of years spent in full-time education and DESI. 108

5.1 Conceptual model 123
5.2 Distribution of the dependent variables (N = 48,047). Source: EVS (2020), own calculations. 127

5.3 (a) Distribution of the change in KOF Globalization Index between 2007 and 2017. Source: KOF Globalization Index (1970-2020), own elaboration (b) Distribution of the NRI across countries. Source: World Economic Forum (2016) 130
5.4 Average acceptance of surveillance by country with 95% Confidence intervals. Source: EVS (2020). Own calculations. 133
5.5 Predicted Acceptance of Surveillance by education and change in KOF Globalization Index (based on M6 in Table 5.2). Source: EVS (2020). Own calculations. 139
5.6 Predicted Acceptance of Surveillance by education and NRI (based on M7 in Table 5.2). Source: EVS (2020). Own calculations. 140

B.1 Frequency distribution of willingness to use the COVID-19 HP, N = 1,454. 167
B.2 Example of vignette. 171
B.3 Correlation matrix of variables included in the regression models. 174

C.1 Proportion of non–internet users by country. The dotted line represents the grand mean across countries. 180
C.2 Predicted values of non-internet use and 95% confidence intervals by 10th, 50th and 90th percentile of years spent in full–time education and DESI. 181

D.1 Distribution of the ES-ISCED educational attainment by countries (N = 48,047) Source: EVS (2020), own calculations. 186
D.2 Average score on item measuring the preference for a strong leader across countries with 95% confidence intervals. Source: EVS (2020), own calculations. 187
D.3 Average score of institutional knowledge across countries with 95% confidence intervals. Source: EVS (2020), own calculations. 190

D.4 Predicted Acceptance of Surveillance by education (based on M1 in Table 5.1). Source: EVS (2020). Own calculations 191
D.5 Fixed and random slopes (estimated from M5, Table 5.2). Dashed lines represent the slopes of education for the pooled sample whereas the solid lines represent the slopes for each country. Source: EVS (2020). Own calculations. 192

List of Tables

2.1 Factor loadings and uniqueness after principal components factor analysis (N = 1,504). 42
2.2 Ordinal logistic regression of trust in social media on relevant covariates. 45
2.3 Cross Tabulation of Trust in Social Media in the pre-test and in the post-test. 47
2.4 Multinomial logistic regression of change in trust in social media (Reference = No Change). 49

3.1 Summary of experimental conditions and hypotheses. 70
3.2 Descriptive statistics. 72
3.3 PPO models of Willingness to use the COVID-19 HP by individual characteristics and experimental conditions. 74

4.1 Factor loadings and uniqueness of principal components factor analysis (N = 21,450). 95
4.2 Descriptive statistics of individual characteristics (N = 21,177). . 98
4.3 Multilevel linear regression analyses of e-privacy management on individual characteristics (N=21,177). 104
4.4 Multilevel linear regression analyses of e-privacy management on individual and country characteristics (N=21,177). 107

5.1 Multivariate multilevel linear regression of AoS on individual characteristics (N=48,047 in 33 countries). 134
5.2 Multivariate multilevel linear regression of AoS on individual and country characteristics (N=48,047 in 33 countries). 137

A.1 Descriptives of all trust-items. 160
A.2 Descriptives of the continuous variables. 161
A.3 Descriptives of the categorical variables. 162
A.4 Ordinal logistic regression of trust in social media on relevant covariates, including trust in government instead of institutional trust. 163
A.5 Multinomial logistic regression of change in trust in social media (Reference = No Change), using trust in government instead of institutional trust. 165

B.1 Testing the proportional odds assumption. 168
B.2 Question wording and answer categories per variable. 172
B.3 Partial proportional odds models of Willingness to use the COVID-19 HP by individual characteristics and experimental conditions, with interaction terms. 176

C.1 Multilevel linear probability model of non-internet use on individual characteristics. 183
C.2 Multilevel linear regression analyses of e-privacy management on individual characteristics. 184

D.1 Percentage of missing values (Don't know + I prefer not to answer) per variable. 188
D.2 Descriptive statistics of individual-level variables. 188
D.3 Factor loadings after promax rotation (N = 54,689). 189

To G.,
'cause this gift will last forever

Ron Swanson: "APRIL! Listen I was trying to buy this hand crafted mahogany wood model of a B-25 Mitchell Panchito aircraft [...] I went to this website and this ad popped up, 'Hey Ron Swanson, check out this great offer.' "
April Ludgate: "What's your question?"
RS: "My question is, what the hell?"
AL: "Like how do they know who you are?"
RS: "Yeah,"
AL: "OK, um, there are these things called cookies, where like if you go to a site and buy something it will remember you and create ads for other stuff you might wanna buy."
RS: "So it learns information about me? Seems like an invasion of privacy."
AL: "Dude, if you think that's bad, go to Google Earth and type in your address."
(Ron takes his computer to the dumpster)

Ron Swanson on Parks & Recreation, season 4 episode 9 "The Trial of Leslie Knope", 2011 - watch the scene at https://edu.nl/3ycjt.

Introduction

I • DATAFICATION AND RISKS

The expansion of ICTs has tremendously enhanced life opportunities: an enormous amount of information is available to citizens, governments, and firms on all corners of the world, which use it in combination with powerful digital tools to make processes more efficient, and – in many cases - safer. In recent years, the advent of Big Data has nurtured the process of datafication of society, or 'the process by which subjects, objects, and practices are transformed into digital data' (Southerton, 2020, p. 1), which now involves virtually every aspect of social life. Think of the quantification of friendships enabled by Facebook via likes and comments, but also –- in the framework of the COVID-19 pandemic – the quantification of exposure risk via the collection of information on social encounters via digital contact tracing.

In current understandings, datafication consist of two elements (Southerton, 2020): the translation of social/human life into machine-readable data, and 'the generation of different kinds of value from data' (Mejias & Couldry, 2019, p. 3). The first element, i.e. rendering social life into quantified bits, has to do with a general tendency to quantification common to all modern societies (cf. Mennicken & Espeland, 2019). Nowadays, however, it is boosted by inexpensive data storage solutions, powerful machine learning tools, and algorithms, that enable to process and analyze large volumes of data at relatively little cost and effort. In this sense, datafication has led to the quantification of phenomena previously uncountable (Cukier & Mayer-Schoenberger, 2013), an endeavor which surely nurtures cumulation of knowledge and heightened efficiency in many aspects of life. For instance, the quantification of health-related behaviors enabled by smartphones and smartwatches nowadays can lead to an earlier diagnosis of health conditions, thus potentially improving the quality of life.

The second element of datafication, i.e. value generation, has to do with the monetization of the collected data (Mejias & Couldry, 2019), a mechanism which is well incorporated into economic dynamics. Internet-based firms, such as Google and Facebook, thrive amidst the emergence of a new logic of capitalist accumulation which Zuboff labelled 'surveillance capitalism', i.e. a 'new form of information capitalism [which] aims to predict and modify human behavior as a means to produce revenue and market control' (Zuboff, 2015, p. 75). It is through the logic of surveillance capitalism that computer-mediated interactions of any kind are put to new uses, including the extraction and analysis of large volumes of data, which can be then accessed behind compensation by third parties. For instance, most social networking sites base their business models on gathering their users' information and preferences to sell these as data for targeted advertising (Zuboff, 2015).

The transformation of social life into data often occurs via cookies, files that record information about an individual's online behavior and transfer it to the website provider. The question included in the title of this thesis, 'Do you want cookies?', refers to the idea that cookies always had this double meaning of being something allegedly harmless, but that can also be used to trick someone into dangerous situations. Children are socialized not to accept cookies (or sweeties) from strangers not to make themselves vulnerable; yet, adults nowadays automatically accept privacy policies and consent to digital tracing via cookies, exposing themselves to datafication processes, as painfully learnt by the very private fictional character Ron Swanson when a banner shows his name (see page xxv) because a website, via cookies, has learnt information about him.

Whereas initially this process of datafication was met with techno-enthusiasm (see Cukier & Mayer-Schoenberger, 2013), critics now illustrate how the datafication process not only creates new uncertainties, but also enables new ways in which old uncertainties are reproduced (Southerton, 2020; van Dijck, 2014). For instance, datafication is accompanied by a shift towards dataveillance, that is 'the continuous surveillance through the use of (meta)data' (van Dijck, 2014, p. 198). Unlike surveillance, dataveillance concerns the collection of data independently from specific purposes (van Dijck, 2014), and relies on the collection of metadata, i.e. data about data, and 'data exhaust', i.e. the byproducts of users' computer-mediated activities. The data extracted from an online activity does not concern only the content of that activity, but also the contextual information. To exemplify, when uploading a picture on Instagram, the relevant information which can be recorded is not (only) the content of the picture – e.g. who was in it, what was represented, where it was taken -- but also a long series of hidden information, such as: which phone model was used to take the picture, which kind of internet connection was used when uploading it, which is the internet provider. These vast amounts of data allow profiling individuals and automatically classifying them in order to determine 'who should be targeted for special treatment, suspicion, eligibility, inclusion, access, and so on' (Lyon, 2005, p. 20). In other words, generated information about individuals is used as a social sorting mechanism which affects individual opportunities and discriminates, for instance by denying access to credit (Lupton, 2016; Lyon, 2005; Mann & Matzner, 2019). The literature provides many examples of how social imbalances are embedded in data and technologies (Joyce et al., 2021), and how data-driven systems generate discriminatory mechanisms against the most vulnerable strata of the population (Brayne, 2017; Eubanks, 2018; S. Park & Humphry, 2019).

The implication for individuals is that while the datafied society requires a constant production of data about the self, it also generates risks of disclosure of information which may not be meant to be publicly available, and with unforeseeable consequences for the future use of these pieces of information. It is important to understand how people react to these datafication-induced consequences, since this has broad implications not only on the ability to seize the opportunities of the digital society, but also on the legitimacy of governmental policies which are increasingly faced with decisions concerning the deployment of digital tools, e.g. technology-based surveillance. A sociological perspective, with its integration of societal processes with individual chances and opportunity appears necessary to contribute to understanding the challenges that datafication poses nowadays, and the way people react to them.

In the remainder of this chapter, I contribute to this endeavor by looking at datafication through the concepts of the risk society framework proposed by Beck (1992). The link between the risk society and datafication -- to date -- has not been explicitly addressed in previous literature, though some elements can be traced in Lupton's attempt to sketch the main features of a 'Digital risk society' (Lupton, 2016). Beck's risk society framework provides a lucid account of the role of technological progress in contemporary societies with its contradictions, and can thus be helpful to shed light on datafication processes. At the same time, the process of datafication constitutes an interesting case to empirically test the validity of the reflexive modernization thesis, thus tackling one of the most prominent critiques to Beck's framework, i.e. the lack of empirical grounding (Burgess et al., 2018; Mythen, 2004, p.71). Section 2 introduces an overview of the main concepts of the risk society framework. In section 3, I incorporate the risk society perspective to issues arising from datafication, and identify the gaps that the empirical chapters of this thesis will tackle.

2 • THE RISK SOCIETY

In *The Risk Society: towards a second modernity*, Beck (1992) describes the shift from the classic industrial society to a new form of society, the risk society. Whereas the former is based on the distribution of wealth, the latter is concerned with the distribution of risks. Risks are preoccupations about the future (Giddens, 1999) and can be defined as 'believed expectations of catastrophes' (Wimmer & Quandt, 2006, p. 341); particularly relevant are manufactured risks, which are induced by scientific and technological progress (Giddens, 1999), among the forces that make modernization itself. In other words, in an attempt to gain control against natural hazards, modern science and technology threaten the existence of individuals by producing man-made consequences, unlike pre-industrial dangers which were considered as given and stemming from the outside (Elliott, 2002; Giddens, 1999). Prominent examples of manufactured risks are climate change caused by human activity and bacterial antibiotic resistance caused by excessive use of antibiotics (Burgess et al., 2018), but also digital freedom risks caused by extensive surveillance online (Beck, 2013).

Central to Beck's risk society theory is the concept of reflexive modernization (Beck, 1992), a process by which modernity starts to question its own advancements, or 'the modernization of modern society' (Beck et al., 2003, p. 1). Rather than following a path of gradual evolution and expansion, reflexive modernization 'refers to a boomerang effect, where mostly unplanned results of (production) processes in modern societies backfire on these societies and force them to change' (Wimmer & Quandt, 2006, p. 337). Reflexivity does not mean heightened consciousness generally, but rather a heightened consciousness of the limits of modernization (Latour, 2003, p. 36). With reflexivity, the core elements of modernization – e.g. the nation-state and technological progress – are constantly questioned (Beck et al., 2003; Lupton, 1997; Wimmer & Quandt, 2006).

Manufactured risks tend to be invisible to human perception (Beck, 1992, 2013), because they often do not hurt physically and because they require casual interpretations which must be 'implied to be true, believed' (Beck, 1992, p. 28). This raised the question of how individuals perceive them. Before Beck, the mainstream explanation of risk perception in the social sciences was the cultural explanation (Douglas & Wildavsky, 1983). This perspective sees risk as socially constructed and explains that people assign different weight to different types of risk depending on their worldviews or, more generally, ideologies (Knight & Warland, 2005; Wildavsky & Dake, 1990). Instead, as explained by Burgess et al. (2018), according to Beck 'risk is very clearly regarded as an idea in its own right relatively independent of the hazard to which it relates' (Burgess et al., 2018, p. 2), feature which also makes risks often difficult to calculate. The latency of such risks can thus only be broken by knowledge which, however, exposes risks to a process of social definition and construction in an authentic dynamic of reflexive modernity (Beck, 1992): while scientific knowledge is essential to identify risks – and solutions - in contemporary societies, science and technologies are also increasingly targeted by criticism and doubt (Mythen, 2004, p. 59). Knowledge, in Beck's perspective, hence becomes a prevalent element in the stratification mechanisms amidst a process of general individualization of life-chances, and produces the conflict between those who profit from risks and those who are afflicted by them- which is also the conflict between those who define risks and those who consume risk definitions (Beck, 1992, p. 46).

There are many actors and institutions involved in the process of social definition of risk: the predominant force being science, devoted to structure knowledge; there are also mass media in charge of dissemination, and companies and politics responsible of taking decisions on the acceptable levels of risk (Beck, 1992). This is exemplified by the concept of *relations of definition*, which is 'a panoply of institutions and agencies involved in the uncovering and communication of risk' (Mythen, 2004, p. 54). However, the institutions which would be responsible of shielding from such risks are intertwined with the same dynamics that produced the risk (Burgess et al., 2018), leading them to deflect responsibility in order to avoid delegitimization (Mythen, 2004). This process is labelled by Beck as *organized irresponsibility* and 'refers to the way in which institutions are forced to recognise catastrophic risks whilst simultaneously refuting and deflecting public concerns' (Mythen, 2004, p. 60).

Manufactured risks are also global in nature, which – however - does not mean that risks are equally spread across regions in the world (Beck, 2002): for instance, each country experiences the consequences of climate change in some form, but this is not to deny that some regions are hit worse than others (Beck, 2002). The pressing nature of material needs in peripheral regions of the world leads to the suppression of intangible hazards, or in Beck's words 'in the competition between the visible threat of death from hunger and the invisible threat of death from toxic chemicals, the evident fight against material misery is victorious' (Beck, 1992, p. 42). This point underscores the importance of the contextual conditions which enable the acknowledgment of risks in a given country or region. For instance, admitting to the deadly nature of some chemical industrial processes may be more difficult in the countries which host polluting plants in exchange for employment and better living conditions of their citizens.

3 • THE CONTRIBUTION OF THE THESIS: UNDERSTANDING THE DATAFIED RISK SOCIETY

At the core of the Risk society is the observation that 'the speeding up of modernization has produced a gulf between the world of quantifiable risk in which we think and act, and the world of non-quantifiable insecurities that we are creating' (Beck, 1992, p. 40), and datafication is very well engrained in this dynamic. On the one hand, datafication is sustained by a general tendency to quantifying, automating, and rendering elements analyzable, process which nurtures the modernization tendencies of instrumental control (Wimmer & Quandt, 2006, p. 345): after all, the continuous gathering of data and information allows to better know phenomena, detect anomalies, and predict dangers. In a reflexive dynamic, however, this generates unintended consequences and insecurities – or better, manufactured risks – whose definition depends on the interplay between different social actors and institutions, i.e. the relations of definitions , and responsibilities are often deflected in an attempt of organized irresponsibility.

Beck himself, for instance, described the 'digital freedom risk' as the global risk derived from extensive state surveillance enhanced by modern digital technologies which threatens our ability to control the flow of information about ourselves (Beck, 2013). In other words, powerful digital surveillance technologies nowadays hinder the ability to regulate the exchange of information between the individual and the social surroundings, i.e. one's privacy. These information flows are an essential part of communication, and are necessary to maintain interpersonal relationships (Anthony et al., 2017). Yet, in a datafied society, individuals are often denied the possibility to decide what to reveal and what to conceal about themselves, for instance because some information can be inferred from metadata even when not provided explicitly. While this is not something completely new – state surveillance has achieved similar goals in many instances in recent decades - the pervasiveness of these processes and the involvement of different types of institutions (e.g. private companies) alongside public authorities is unprecedented.

The latency of manufactured risks is one of the pivotal points of Beck's risk society thesis (Beck, 1992), the idea being that risks caused by modernization tend to be invisible. This is also the case for datafication-induced risks: risks like digital freedom risks (cf. Beck, 2013) and social sorting (Lyon, 2005; Mann & Matzner, 2019) in most cases do not hurt physically nor directly. When citizens are affected by these negative outcomes, it is difficult to attribute them to datafication processes, for at least two reasons. First, because in a typical risk-society dynamic of organized irresponsibility (cf. Mythen, 2004, p. 61) institutions carrying out datafication activities tend to try to dodge responsibilities by alimenting a narrative of 'data neutrality' (cf. Boyd & Crawford, 2012; Joyce et al., 2021) and 'algorithmic unbiasedness' to maintain their legitimacy. Second, because it takes specialistic knowledge to be able to draw casual links between phenomena (Beck, 1992). The next two sections describe these two points in larger details, and identify the gaps to be tackled by the remainder of the thesis.

Trust and organized irresponsibility

The complexity of modernity requires reliance upon expert systems which are inescapable because they produce and enact knowledge which is otherwise not accessible by ordinary citizens (Giddens, 1990). This requires trust, which is 'usually routinely incorporated into the continuity of day-to-day activities' (Giddens, 1990, p. 90). Yet, confronted with the fallibility of experts, individuals start to question expert systems and, therefore, trust should not be taken for granted yet 'worked and won' (Meyer et al., 2008). Similarly, in the risk society framework as formulated by Beck, trust in modern institutions is expected to drop as consciousness about their limitations grows in a reflexive dynamic (Beck, 1992). This extends to different types of institutions: for instance, the democratic deficit model by Norris (2011) postulates that as expectations on the role of government rise, the limitations of its potential achievements become more visible, resulting in a drop in its perceived legitimacy. Hence, while more and more trust is fundamental to the functioning of complex societies increasingly relying on automated abstract systems, trust is also threatened by the reflexive discovery of the inherent limitations of these systems.

Despite a wealth of empirical research on trust, the link between Beck's risk society and trust is not adequately addressed. Some studies use trust in science as a proxy of reflexive mindset at the level of individuals (cf. De Keere, 2010; Price & Peterson, 2016), however this does not fully shed light on trust dynamics within a complex systems of relations of definitions. The reflexive modernization perspective has been applied to studies on trust in the health care realm (cf., e.g., Lupton, 1997; Ward, 2006), but it is a peculiar case as trust in a system (health care) passes through micro-level interactions between social actors (e.g. patient-doctor interaction). In the case of datafication, there is hardly any face-to-face interaction which provides an experiential basis to be generalized to a broader system. Therefore, one of the contributions of this thesis is to tackle the following question: *to what extent does institutional trust drop when the inherent limits of datafication processes become visible?*

The awareness over manufactured risks jeopardizes the legitimacy of institutions which are involved in the process of definition of those risks, and which therefore react by trying to deflect responsibilities, i.e. by enacting organized irresponsibility to try to delay the uncovering of risks. This response is articulated differently by different types of institutions. Indeed, as seen above, the process of social definition of risks – including the datafication-induced risks – depends on multiple institutions and social actors, i.e. the relations of definitions (Mythen, 2004). In the case of datafication, the pool of institutions involved includes not only 'traditional' institutions and agencies (e.g. governments, media, academia), but also a constellation of private institutions, in particular tech companies and social media platforms (cf. van Dijck, 2014, p. 204).

To give but one example of organized irresponsibility in a tech firm, in 2018 an article in the Washington Post[1] and a blogpost on Gizmodo[2] raised

[1] See https://www.washingtonpost.com/technology/2018/11/16/wanted-perfect-babysitter-must-pass-ai-scan-respect-attitude/ (Accessed on 17-12-2021).
[2] See https://gizmodo.com/predictim-claims-its-ai-can-flag-risky-babysitters-so-1830913997 (Accessed on 17-12-2021).

awareness on Predictim, an AI system built to scan the social media activities of potential babysitters to determine their 'risk level'. The author of the Gizmodo post found that dark skin people would get consistently higher risk scores than non-dark skin people, thus asking to the developers of the app whether it was possible that some racial bias had crept in their AI system. Reportedly, the CTO of the company deflected responsibility by stating the following:

> I can guarantee you 100 percent there was no bias that went into those posts being flagged. We don't look at skin color, we don't look at ethnicity, those aren't even algorithmic inputs. There's no way for us to enter that into the algorithm itself. (Merchant, 2018)

A clear underestimation of at least two problems of the datafied society is visible in the statement: first, even if ethnicity is not explicitly fed into the AI system, there are other information strongly correlated with it (e.g. language used in the posts) which may end up influencing the final result on the implicit basis of ethnicity. Second, if people of a certain ethnicity are labelled disproportionately as higher risk in the training data the AI system learns on – training data which are often compiled by humans - that bias will also be learnt by the AI system, reinforcing inequalities (Joyce et al., 2021). Denying these findings by presenting the AI system as neutral, as attempted by the CEO of Predictim, can be seen as a manifestation of organized irresponsibility.

When implemented by public institutions, datafication processes produce risks for the monitored subjects whose opportunities (e.g. access to welfare benefits, physical safety) are contingent upon information generated about themselves; yet, these risks are masked by narratives of efficiency improvements and unbiasedness. Indeed, as showed by Brayne (2017), the use of predictive analytics for policing is promoted as a way to protect against crime while relying on actual facts/data, rather than on the subjective prejudice of police officers.

Similarly, the Australian government introduced AI systems in support of the debt recovery system and the national insurance agency in an attempt increase efficiency by saving costs and increase accessibility (S. Park & Humphry, 2019). In these ways, civil servants and police officers are set free from the responsibility for the potential relentlessness against specific social groups, which occurs anyway due to the reproduction of social imbalances in data processes (Joyce et al., 2021) but is now justified because based on a neutral analysis of data (Brayne, 2017; S. Park & Humphry, 2019).

The effectiveness of the institutional attempts of organized irresponsibility to maintain risks latent, however, has not been systematically addressed empirically in the literature thus far, despite constituting an important question for the future of societies. Therefore, in the thesis, I ask *to what extent are individuals aware of the manufactured privacy risks generated by datafication processes?* As citizens will be increasingly confronted with the deployment of these technologies, awareness is needed for them to hold the institutions accountable for the outputs of automated data-driven processes. Otherwise, if attempts of organized irresponsibility are successful, the manufactured privacy risks deriving from datafication processes may silently escalate.

To answer these questions, two empirical chapters deal with specific events which, in different ways, are characterized by latent risks generated by institutions. In the first study, I explore whether the uncovering of privacy risks stemming from social media has an impact on the trustworthiness of social media themselves. In the second study, latent privacy risks and visible health risks provide a reference frame for the acceptability of a new, potentially invasive, technology proposed by the government. The Netherlands is used as a setting for these studies: as a country, it can be considered well into reflexive modernity dynamics as it is confronted on a daily basis with the risks deriving from human activities, such as earthquakes induced by gas extraction and rising sea level due to climate change. Additionally, the Netherlands has a high level of penetration of ICTs[3], which makes the country well engrained in datafication dynamics.

Chapter 2 presents a study on the alleged drop in trust in social media as a result of the uncovering of the Cambridge Analytica scandal and of a more general debate around online privacy. With the Cambridge Analytica scandal, it became visible how social media (and other companies) were able to harvest information about their users and monetize them. The idea behind the study is that once it becomes known that a platform such as Facebook can sell data to third parties for opaque purposes, it constitutes a breach of trust, and individuals should become more skeptical of social media: in a reflexive dynamic, respondent should see the limitations of transferring social life online, where it can be datafied and lead to unforeseen consequences. In this context, the 'worked and won' dynamic of trust is linked to the endogenous account of institutional trust (e.g. institutional trust is based on a critical evaluation of the performance of the institution under scrutiny) which, in the chapter, is compared against the exogenous explanation (e.g. institutional trust is rooted in cultural and personal dispositions).

Chapter 3 explores the acceptability of the COVID-19 health pass, a potentially privacy-invasive data-based technological solution to the spread of the COVID-19 pandemic, among Dutch respondents. In coarse terms, the investigation concerns the acceptability of a latent risk (i.e. the disclosure of sensitive information via the health pass) against the backdrop of the not-so-latent-risks deriving from resuming social life amidst a pandemic. The contextual integrity framework (Nissenbaum, 2010) is hereby used to understand how different features of the health pass may affect the perception of the risk of disclosure, alongside the cultural predispositions of respondents.

Due to the unavailability of datasets on the specific topics, both Chapter 2 and Chapter 3 are based on data specifically collected via a probability-based

[3] See, e.g., https://digital-strategy.ec.europa.eu/en/policies/desi-netherlands.

online panel in the Netherlands (LISS panel), using questionnaires designed by the authors of the studies and described in larger details in the respective chapters[4]. The LISS panel provides the unique opportunity to collect data among representative samples of Dutch households maintaining high scientific standards, e.g. probabilistic sampling and targeted actions to reduce coverage bias.

Knowledge as a risk stratification mechanism

Provided that in Beck's perspective manufactured risks only exist in so far as knowledge about them exists, the second contribution of this thesis is to apply the risk society perspective to investigate knowledge asymmetries in recognizing datafication-induced risks and breaking their latency. This endeavor allows identifying sources of inequalities as some individuals may be better able to seize the opportunities offered by the datafied society thanks to their consciousness of the risks induced by it.

Central to Beck's thesis is that risks are global in nature, yet some individuals are more affected than others, though this follows different fault lines than the class divisions typical of the industrial society according to the author (Beck, 1992). Indeed, in his original formulation, Beck claimed that in the risk society the relevance of class as an element of stratification diminishes[5] and that risk is

[4] More information about the LISS panel can be found at: www.lissdata.nl

[5] The claim on the reduced relevance of class is not unchallenged; in particular Curran (2013) reported many critiques to this approach, and showed how class becomes even more important as a stratification factor for the distribution of risks. While allowing Beck's idea that 'the initial distribution of risk may be egalitarian, as individuals become increasingly cognizant of the effects of these different risks and their relative distribution, the ability to escape the sources of [...] risk will likewise be highly differentiated' (Curran, 2013, p. 56). After all, economic power – which is strongly related to class - helps reducing the exposure to risks, once risks are recognized thanks to knowledge – which is again correlated with class. For

egalitarian (Beck, 1992). Accordingly, dangers in the classic industrial society were mostly concentrated among the low social classes, whereas modern risks potentially threaten everyone because they are global in nature, and not necessarily bound by space and time constraints; yet, only some individuals know about risks and hence face them (Beck, 1992). In Beck's words, 'this transmission through knowledge means that those groups that tend to be afflicted are better educated and actively inform themselves. [...] risk consciousness and activism are more likely to occur where the direct pressure to make a living has been relaxed or broken, that is, among the wealthier and more protected groups (and countries)' (Beck, 1992, p. 53). According to Beck, those who have better education, more knowledge and/or more access to information are better able to become aware of risks and, hence, be afflicted by them.

INFORMATION, KNOWLEDGE, EDUCATION

Before proceeding, it is important to specify the distinction between knowledge, education and information. Despite the differences in conceptualization, the three are strongly tied and there is some overlap. Thanks to the advent of mass media first and the internet then, access to information is becoming more and more affordable for different social strata and areas of the world. Even though the share of population online has grown tremendously on a global scale between the 90s and more recent days, the growth has not been homogeneous. To date, there are large differences in the rate of access to the internet in different countries, with developing countries showing a lower share of internet users compared to more affluent countries (see Figure 1.1).

At the level of individuals, extensive research has been done on digital divides and on the way access to information technologies reproduces offline inequalities along the lines of gender, race, social class, and educational level (Hargittai,

instance, by affording a house in a better position, individuals are better shielded by pollution, net of the fact that polluted air can potentially reach everyone.

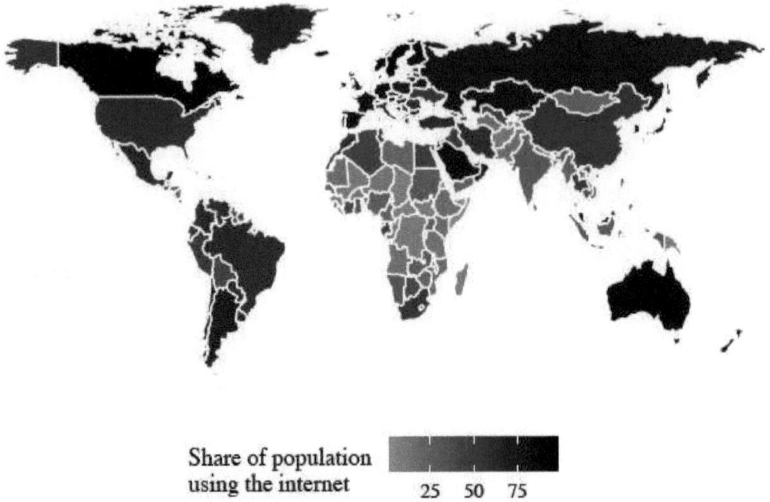

FIGURE I.1 Share of internet users by country between 2015 and 2017. Own elaboration of data from Ourworldindata.org.

2002; Helsper & Reisdorf, 2017; Scheerder et al., 2017; van Deursen et al., 2011; van Dijk, 2005, 2013). Even once diverse people are online, research has repeatedly showed the existence of knowledge gaps, i.e. the enhanced opportunities of those with a higher social status to acquire information from the internet (Bonfadelli, 2002; Gerosa et al., 2021; Lind & Boomgaarden, 2019). Access to and collection of information hence appears to be stratified on one's knowledge and education.

The distinction between knowledge and education is perhaps more blurred. Education can be seen as a formal process of acquisition of knowledge; knowledge, however, can be gained also outside the educational path and in more informal settings. For the purposes of this thesis, the focus will be on education, though occasionally some forms of informal knowledge will also be addressed. The advantages of focusing on education are manifold: first, educational attainment, as in the highest level of education achieved, indicates a structural position within society and, even if it does not always directly quantify the amount of knowledge an individual has, it has a signaling power which enhances life-chances, labor market outcomes and the social status of individual (Bol & van de Werfhorst, 2011; Bovens & Wille, 2017). Second, the measurement of knowledge is complex and domain-specific, whereas the level of education can be assessed with general measures which are also validated in comparative perspective, and is therefore more suitable for investigations among the general population. Finally, shedding light on the role of education calls into play the role of public institutions responsible of its provision, whereas informal knowledge is relatively more dependent on individual initiative.

EDUCATIONAL GAPS, IN CONTEXT

As said above, an essential feature to recognize manufactured risks is to be able to draw causal interpretations – that is, to be able to recognize the risk as induced by modernization itself (Beck, 1992). This requires some degree of skepticism towards scientific and technological progress and institutions (De Keere, 2010), which can be called reflexive mindset (Achterberg et al., 2017) to mimic the upper-level dynamic of reflexive modernization, and which is expected to be found among those with a higher level of education. Empirical evidence on this is mixed: at the individual level, education has been found to be positively correlated with trust in science against the expectations deriving from the theory (Achterberg et al., 2017; De Keere, 2010; Price & Peterson, 2016); other studies, however, suggest that material deprivation (which is strongly tied to achieving a low level of education) hinders reflexivity (Ward, 2006; Ward & Coates, 2006). This inconsistency in the empirical findings leaves an open question as to *whether an individual's educational achievement affects their ability to acknowledge datafication-induced risks through a reflexive mindset, or lack thereof.*

However, the ability to acknowledge manufactured risks does not solely depend on individuals, since contextual constraints (e.g. the facility to access information, or the transparency of institutions) also enable, or hinder, the process. For instance, Price and Peterson (2016) found that confidence in science is lower in countries with higher rates of university enrollments and internet access, despite the lack of association between education and trust in science at the individual level. In other words, in countries with larger availability and production of knowledge overall, science was found to be looked at with more skepticism. This finding underscores the importance of considering the layer of differences across national contexts: not only may they explain the uneven flourishing of a reflexive attitude towards modern institutions, but they may also affect the size of the educational differences in the acknowledgement of manufactured risks. The level of digitalization in a given country is an important aspect that bridges datafication processes to the risk society perspective. The availability of ICTs and digital tools is a pre-requisite of contemporary datafication process, which require complex and powerful computing capacity. However, as seen above, due to reflexive modernization, not only the expansion of these technologies and datafication processes brings about manufactured risks, but also the availability of and easier access to information to a wider audience actually nurtures reflexivity itself. There are several implications. First, the higher the level of digitalization of a country, the more datafication-induced manufactured risks are generated. Second, the opportunity of individuals to develop a reflexive mindset is enhanced by the accessibility of information. For these reasons, the final question driving this thesis can hence be formulated as follows: *to what extent do educational gaps in the acknowledgement of datafication-induced risk vary by the levels of digitalization in different European countries?*

To address these two questions, two empirical chapters focus on the role of formal education in acknowledging datafication-induced risks in a cross-national perspective. In both chapters, the validity of expectations derived from the reflexive modernity thesis is tested, and contrasted with alternative theories in order to be able to shed light on the actual mechanisms.

Chapter 4 investigates the educational gap in e-privacy management, and seeks to explain whether and why individuals with a higher level of educational attainment are more prone to manage their privacy online (thus paying more attention to datafication processes). Additionally, the chapter investigates whether the educational gradient becomes larger or narrower depending on the availability of ICTs in a country. In this chapter, as an alternative to the reflexive modernization thesis, expectations deriving from the digital divide and the diffusion of innovation perspective are tested.

Chapter 5 focuses on the educational gradient in acceptance of surveillance, both in online settings and in public areas; it also explores how the educational gradient is conditioned by some contextual characteristics. As an alternative to the reflexive modernization thesis, and given the tension between security and privacy (risks) entailed by surveillance, this chapter looks at the cultural backlash theory (Norris & Inglehart, 2019). This theory predicts a stronger demand for security and hence support for surveillance among vulnerable strata in countries that underwent rapid social and cultural changes.

Chapter 4 and Chapter 5 reuse existing secondary data (respectively, Eurobarometer and European Values Study) which are freely available for scientific purposes on trusted repositories. Both surveys include data from representative samples of multiple European countries. Chapter 4 and Chapter 5 include also contextual-level data about the country of residence of the respondents, with datasets coming from Eurostat, World Bank, and other repositories of aggregate statistics. The use of multilevel regression models in Chapter 4 and Chapter 5 allows to account not only for individual-level dynamics but also to incorporate contextual level characteristics, and even to condition the former on the latter.[6]

[6] The data preparation steps and statistical analyses included in the chapters are conducted using popular statistical software and documented on scripts and syntaxes which are stored in Data packages; Stata 16 was used for the analyses included in Chapter 2 and Chapter 4, whereas the analyses of Chapter 3 and Chapter 5 have been performed on Rstudio. Data package including all the scripts and relevant material are deposited in trusted repositories, and linked in each chapter.

In Zuck We Trust?

The sources of trust in social media in times of data privacy controversies

ABSTRACT

Even though billions of people use social media daily, little is known about the extent people trust them, what explains their trust, and whether particular events related to online data privacy influences trust. Studies on the roots of trust in institutions are divided over whether trust is endogenously explained by the functioning of these institutions, or whether trust in such institutions is exogenous to other factors. In this chapter, we concentrate on the period around the Cambridge Analytica data breach scandal and the introduction of the European General Data Protection Regulation to gain insights on the determinants of trust in social media. To study this, we rely on a unique panel study as part of the Dutch wave of the European Values Study 2017, questioning a representative sample about their trust in social media before and after the controversy over online data privacy. Analyses suggest that trust in social media is distinct from other types of institutional trust and show that the Dutch have a rather low level of trust in social media. Further, trust in social media is strongly affected by cultural explanations, among which postmaterialism and reflexive modernity. The data breach turmoil did not strongly scratch trust, providing mixed support for the endogenous roots of trust. We conclude the chapter with implications for the concept of trust, and its study.

This chapter is a joint work with Tim Reeskens. Replication materials for this chapter are available on a private repository on OSF (https://edu.nl/p7dq8).

1 • INTRODUCTION

Social media like Facebook, Twitter and Instagram have become widespread in everyday life, with approximately 65 percent of the European internet users reported to be active on social media[1]. Social media not only serve the purpose of facilitating interpersonal relationships, but they are important for democratic processes as people increasingly use them for gathering news and information (Dubois & Blank, 2018, p. 733). Thus, although social media platforms are owned by private companies, public values such as transparency and freedom of expression – traditionally anchored to public institutions – are nowadays increasingly reproduced through social media (Gillespie, 2010; van Dijck et al., 2016). This raises important questions on how citizens relate to social media and react to their pitfalls.

Lately, netizens – i.e., citizens of the World Wide Web – have been confronted with major challenges. Not only are social media increasingly under scrutiny for facilitating so-called 'fake news', but privacy issues became manifest. Particularly, a wide debate over online privacy sparked in the spring of 2018. Firstly, in March 2018 major news outlets reported that political consultant firm Cambridge Analytica harvested personal information of millions of Facebook users without their prior consent and used it for targeted political campaigns. Secondly, in May 2018, the European Union enforced its new law to protect consumer data privacy – the General Data Protection Regulation (GDPR) – which should provide Europeans with greater transparency over the data that institutions collect from them online. While empowering citizens by giving them more control over digital information, the GDPR also unveiled how exposed personal data had been until that moment. Following the enforcement of the GDPR, nearly all European netizens were confronted

[1] See http://appsso.eurostat.ec.europa.eu/nui/submitViewTableAction.do (last accessed: 18-12-2020).

with a flood of emails concerning updated privacy settings, and consenting requests concerning the collection of internet cookies in their online activities.

In this context of risk for individual data privacy, the concept of trust is particularly relevant, as also explained in Chapter 1. In the aftermath of the Cambridge Analytica scandal, Facebook CEO Mark Zuckerberg himself apologized, in an interview with CNN, for the 'major breach of trust' (cf. Hall, 2020). As a matter of fact, trust is involved whenever internet users disclose information about themselves, directly (e.g. subscribing to a service) or indirectly (e.g. accepting internet cookies), as they ought to have the expectation that the digital platforms will use their data confidentially. A better understanding of trust in online social media and how it is affected by a privacy breach is thus extremely relevant in contemporary societies.

The literature is divided over the sources of trust, and whether they are endogenous (i.e. trust in a certain institution responds to the evaluation of its functioning) or exogenous (i.e. trust stems from elements outside the functioning of the institution involved) (cf. Mishler & Rose, 2001). The aforementioned events, by putting online privacy under the spotlight, offer the unique opportunity to engage with this debate while at the same time exploring the foundations of trust in a relatively new institution which is deeply embedded in datafication practices, i.e. social media. If the endogenous theory on trust is valid, we can expect that when confronted with events that increase the salience of online data privacy, individuals would turn their back towards data-handling institutions such as social media. Since little is known about trust in social media, we break our empirical inquiry down in several steps. First, we study to what extent trust in social media correlates with trust in other institutions and engage with the scholarly debate on the dimensionality of institutional trust; second we study the individual sources of trust in social media; third, we analyze whether individual trust in online social media has shifted after the turmoil over online data privacy.

We rely on a unique analytical design incorporated in the Dutch implementation of the European Values Study (EVS), including an item on social media in its institutional trust battery. Whereas the main EVS data collection in the Netherlands took place between September 2017 and January 2018, the peculiar design of the survey allowed to re-approach the respondents for a follow-up survey on trust in social media in June 2018, in the aftermath of the controversy over online data privacy sparked by the Cambridge Analytica affair and the introduction of GDPR, enabling the study of changes in trust in social media over time.

The Netherlands provides an interesting context for this study: not only is the country well embedded in reflexive modernization dynamics (cf. Chapter 1, pp. 9–14), but also the high levels of Internet penetration and digital literacy[2] lead to expect that the Dutch are familiar with social media, either as active users or as bystanders. The Cambridge Analytica-scandal was picked up by traditional media, in particular by the popular television show 'Sunday with Lubach' (*Zondag met Lubach*) in which host Arjan Lubach launched the #Bye-ByeFacebook event, inviting the audience to delete their Facebook accounts. The initiative echoed not only on Facebook itself but also on news outlets. On different issues, the television program showed to have an agenda-setting function (Boukes, 2019), suggesting that, likely, most of the Dutch have been exposed to the events relevant to present study. Moreover, a Eurobarometer survey fielded in 2019 showed that the Dutch were among the most knowledgeable in the EU about the GDPR and the rights it grants (European Union, 2019), signaling general exposure to the topic.

The chapter proceeds as follows. In the second section, we introduce the theoretical framework, from which we derive the hypotheses. We approach this study from the institutional trust-literature, and stretch the causes for trust in institutions to the roots of trust in social media. In the third section, we introduce the analytical strategy and the data. The results are discussed in the

[2] See, e.g., https://digital-strategy.ec.europa.eu/en/policies/desi-netherlands.

fourth section. Finally, we conclude the chapter with relating our findings, showing no clear change in trust in social media as a consequence of rock-bottom levels of trust in them, reflections on the study of trust, as well as the future of social media.

2 • THEORETICAL BACKGROUND

Social media and trusting them

As argued by Norris (2011, p. 19), aligning to others (e.g., Newton, 2001, p. 202), trust can be seen as 'a rational or affective belief in the benevolent motivation and performance capacity of another party.' Accordingly, trust is not only a rational consideration of the trusted party (e.g., Hardin, 1993), but it also stems from moral factors going beyond a pure cognitive evaluation. For Levi and Stoker (2000, p. 476), the essence of trust is that an individual is making 'herself vulnerable to another individual, group, or institution that has the capacity to do her harm or betray her'. In this study, the institutions to which individuals make themselves vulnerable to, and whose benevolent motivations and performance capacity are evaluated, are social media.

Originally, social media promoted diversity, freedom of speech, knowledge sharing, etc... and were surrounded by a techno-optimistic attitude (Kidd & McIntosh, 2016). However, social media conglomerates have been soon accused of generating 'filter bubbles' and 'echo chambers', with users only exposed to views similar to the ones they already held (Croteau & Hoynes, 2019; Dubois & Blank, 2018), thus reinforcing existing attitudes whereas at the same time betraying the 'openness' potential of virtual communities. Therefore, the combination of the initially positive aspirations of social media with the disputed technological implementations provide an opportunity to test how the perception of the 'benevolent motivations and performance capacity' (Norris, 2011, p. 19) of social media translates into individual trust.

On endogenous and exogenous explanations of trust in social media

Sociological insights approach the sources of trust in institutions from two perspectives, namely endogenous and exogenous explanations (cf. Mishler & Rose, 2001). The endogenous account nods to the rational choice theory, since trust is conceived as a cognitive evaluation of the performance of the object of trust (Hardin, 2006). Accordingly, institutional trust depends on the (perceived) functioning of the institution itself, and each institution is evaluated independently; the assessed performance of the institution is thus a function of a direct experience with it. Studies found that, among British samples, trust in the internet is largely affected by the degree of experience with the internet itself (Blank & Dutton, 2012; Dutton & Shepherd, 2006). As for social media, insights on trust in Facebook have shown that people form expectations on the social network's functioning as a technological tool to reach a goal (e.g. connect with friends), but also on its respect of social norms (e.g, it will not be harmful) (Lankton & McKnight, 2011). In other words, when they have to evaluate the trustworthiness of the social medium, respondents value both interpersonal aspects, such as integrity and benevolence (see Mayer et al., 1995), and technology-specific beliefs, such as reliability and helpfulness (Lankton & McKnight, 2011). Social media might comply easily with technological aspirations, as their technical functionalities are constantly refined. Arguably, however, the interpersonal aspects have been challenged by the privacy debate emerged in the spring of 2018. Thus, a straightforward interpretation of the endogenous account is that the data privacy controversy would lead to depressed opinions towards social media.

Studies on the consequences of the Cambridge Analytica turmoil for social media users are scarce. In one of the few attempts, Brown (2020) conducted 10 in depth-interviews among students on the decision to keep or not their Facebook accounts after the Cambridge Analytica scandal. Students reported increased privacy concerns after the scandal, concerns which were, however, either dismissed as distant, or considered a necessary trade-off to be able to use the platform; nevertheless, some users thought that lower engagement with Facebook also meant less privacy to be invaded and adjusted accordingly (Brown, 2020). Eventually, some of the students settled with lower frequency of Facebook use, whereas the need of maintaining social contacts and the role of a Facebook account as access to other services prevented them from completely abandon the medium (Brown, 2020). Such mechanism would explain why a drop in trust could be observed in the first half of 2018 despite the stable if not growing number of Facebook users.

Two related but distinct mechanisms explain why cues of malpractices, amplified by the media as in the case of the Cambridge Analytica scandal and the introduction of the GDPR, should translate into lower trust. On the one hand, there is the concept of agenda setting, i.e. 'a strong correlation between the emphasis that mass media place on certain issues (...) and the importance attributed to these issues by mass audiences' (Scheufele & Tewksbury, 2007, p. 11). On the other hand, priming 'occurs when news content suggests to news audiences that they ought to use specific issues as benchmarks for evaluating the performance of leaders and governments' (Scheufele & Tewksbury, 2007, p. 11). Combined, it is quite plausible that news concerning Cambridge Analytica's opaque practices and the flood of requests concerning renewed access to personal data following the GDPR enforcement led people to attribute more importance to potential privacy violations committed by social media, hence negatively influencing trust in social media themselves.

Opposite to the endogenous sources to institutional trust, exogenous factors have shown to be relevant for explaining variations in trust, too. Accordingly, institutional trust would be mostly explained by personal dispositions and cultural factors. First of all, it is important to consider the association among types of trust. On the one hand, the trust-nexus hypothesis (Hanitzsch et al., 2018) posits that the linkage that ties individuals to social media works akin to other vertical links between individuals and political institutions. As such, it can be expected that trust in social media shares communality with trust in other institutions; a more radical interpretation would be that institutions are all evaluated as one unidimensional object, without differentiation between one institution and the other. On the other hand, studies on horizontal conceptions of social trust predict a spillover-effect (Newton & Zmerli, 2011): social trust should pour into trust in institutions (van der Meer, 2003), including social media. Uslaner (2002), for instance, reported that trusting people displayed a more positive view of the internet, whereas people with low trust have more fear and concerns over their privacy.

Second, trust in social media is expected to be rooted in cultural explanations. The theory of post-materialism by Inglehart (1977, 1997) posits that rising economic and physical security over the past decades caused a 'silent' shift from materialistic values, related to material life conditions, to post-materialistic values, concerned with autonomy and self-expression. Post-materialist individuals should hence hold higher aspirations concerning the potential benefits of social media, but would be at the same time also more skeptical about their functioning (cf. Norris, 2011). Despite their attitudes supporting democracy, post-materialists tend to be more skeptical towards authorities and institutions (Catterberg & Moreno, 2005; Tsfati & Ariely, 2014). At the macro-level, for instance, post-materialism tends to correlate with lower political trust (Inglehart, 1997; Tsfati & Ariely, 2014). Moreover, as elaborated in Chapter 1, the theory of reflexive modernization (cf. Beck, 1992) suggests that higher educated individuals living in post-industrial societies are more aware of the risks entailed by the modernization processes in the contemporary world (Makarovs & Achterberg, 2017; Price & Peterson, 2016), and this would translate into being more knowledgeable about the pitfalls of social media too, hence fueling their mistrust towards social media.

Towards an empirical strategy

To test the relative weight of endogenous and exogenous determinants of trust, we develop an incremental empirical research strategy that addresses the theoretical assumptions using data from the recent wave of the Dutch implementation of the EVS, fielded in 2017.

In the first place, this study will investigate whether trust in social media is conceptually different from trust in other institutions, tackling a core debate in the literature on the dimensionality of trust in institutions.[3] Some, aligning to the exogenous perspective, argue that institutional trust is a unidimensional latent construct because individuals are cognitive lazy, unable to distinguish several institutions, and therefore infer from the general quality of institutions (e.g. Hetherington, 1998; Hooghe, 2011). Others, embracing the endogenous perspective, argue instead that trust in institutions is multidimensional and is a rational response to experiences with the institutions or actors evaluated (e.g. Fisher et al., 2010; Rothstein & Stolle, 2008). In order to test empirically the endogenous roots of trust in social media, the expectation is that social media are evaluated distinctly from other institutions, flowing into Hypothesis 1: *In a test of the dimensionality of trust in institutions, trust in social media appears conceptually distinct from trust in other institutions.*

In a second stage, we identify the profile of people trusting social media by testing several factors that are deemed important. First, related to endogenous explanations, the use of social media is likely to affect the evaluation of their own trustworthiness. Blank and Dutton (2012) found that people became more trusting as the gained more experience with the internet, leading to the expectation that *people who use social media more frequently will also have more trust in them* (Hypothesis 2). Second, turning to exogenous explanations, the

[3] As elaborated further, trust in social media is part of an 18-item battery questioning trust in institutions.

trust-nexus is tested by assessing the correlation between trust in institutions (a vertical relation between individuals and institutions) and trust in social media, *expecting a positive relationship between institutional trust and trust in social media* (Hypothesis 3). Next, the role of horizontal social trust is tested, with the spill-over effect predicting that *social trust is positively related to trust in social media* (Hypothesis 4). Furthermore, cultural explanations are tested in two distinct ways. First, we expect that *postmaterialists will show less trust in social media compared to materialists* (Hypothesis 5) due to their dissatisfaction with institutions (Catterberg & Moreno, 2005; Inglehart, 1997; Tsfati & Ariely, 2014). Finally, we expect the *higher educated to be less trusting of social media than the lower educated* (Hypothesis 6) because they are more likely to perceive the privacy violations of social media as modern risks, following the reflexive modernization thesis (Beck, 1992).

In the last step, the panel structure is exploited to test the viability of the endogenous sources of trust in social media. It will be examined whether, following the turmoil over online data privacy emerged around the Cambridge Analytica scandal and the GDPR-enforcement, peopled adjust their trust in social media. Due to the aforementioned agenda-setting and priming effects, we propose Hypothesis 7: *people trust social media less in response to the data privacy controversy ignited in the spring of 2018.* In a final exploratory step, we also investigate which social groups are more likely to adjust their trust in response to the controversy, since people may differ in their exposure and reactions to privacy issues.

3 • DATA AND METHODS

The European Values Study 2017 Netherlands

To test our hypotheses, we rely on an innovative design implemented in the Dutch wave of the European Values Study 2017 (European Values Study, 2020).

The EVS is a large-scale social survey that takes place in all European countries every nine years since 1981 to investigate (change in) relevant political and social values, attitudes and beliefs. In 2017, the Dutch EVS was collected using a mixed-mode design, with part of the sample interviewed face-to-face (CAPI) and part online (CAWI); only the latter is included in this study. The CAWI survey was integrated into the LISS Panel and administered by CentERdata. Participants in the LISS Panel are selected through a random sample of the population register, and are hence representative for the Dutch population.[4] For the main Dutch CAWI EVS, a matrix design was used in order to reduce the length of the interview and improve response rates. A thorough description of the EVS matrix design can be found in Luijkx et al. (2021).

From September to October 2017, 2,053 questionnaire were returned out of the 2,515 invitations sent (participation rate of 81.6 percent). A second survey to complete the matrix was fielded in January 2018; of the 2,014 invitees, 1,722 completed the questionnaire (participation rate of 85 percent). Throughout the analyses, the matrix design is controlled for by using group-dummies, showing no significant differences between groups (also implying that trust in social media did not change between September-October 2017 and January 2018). This part of the data collection will be called the 'pre-test' and used for the cross-sectional analysis of (1) the relationship between trust in social media and trust in other institutions, and (2) the individual-level sources of trust in social media. After deleting cases with missing data on one or more variables (N=218, or 12.5 percent of the total sample), the pre-test comprises 1,504 respondents.[5]

To investigate whether public opinion regarding trust in social media had shifted due to the data privacy controversy sparked by the Cambridge Analytica affair and the GDPR introduction, in June 2018 1,887 respondents that

[4] To overcome coverage issues, internet access is provided to sampled respondents lacking it.
[5] The pre-test data is publicly available for scientific use, see European Values Study (2020).

participated in the pre-test were invited in a follow-up survey consisting of a few repeated questions from the main EVS. From those invitees, 1,510 questionnaires (80.0 percent participation rate) were returned.[6] Throughout the chapter, we refer to this data collection as 'post-test'. After listwise deletion of missing values from relevant covariates, the sample size is 1,097 respondents.[7] As indicated in Figure 2.1, the interest in the Cambridge Analytica-scandal and the unfolding of the GDPR could be detected by heightened searches over Google.

Measuring Trust in social media

Trust in social media in the EVS 2017 is measured as part of a battery on trust in institutions. The battery is preceded by the question 'Please indicate how much confidence[8] you have in each of the items presented in the next questions. Is it a great deal, quite a lot, not very much or none at all?'. The last item of the list was 'Social media'. Since it might be questioned which social media respondents think of, results from CRONOS[9] showed that -when offered the same question- respondents indicated that they first and foremost think about Facebook, followed by Youtube, Google+ and Twitter (CROss-National Online Survey panel, 2018).

To evaluate whether trust in social media is distinct from trust in other, more

[6] The fact that more respondents have been invited than completed the matrix in January is due to the fact that also respondents who only participated in the first round were invited to participate to the follow-up. This will not hamper the analysis of within-change, as change is calculated on those respondents who answered the trust in social media-question in the pre-test and the post-test.

[7] The post-test dataset is available upon request to the authors.

[8] In Dutch, the word 'vertrouwen' is used, which means both trust and confidence.

[9] CRONOS stands for CROss-National Online Survey Panel, and has been fielded as part of the European Social Survey.

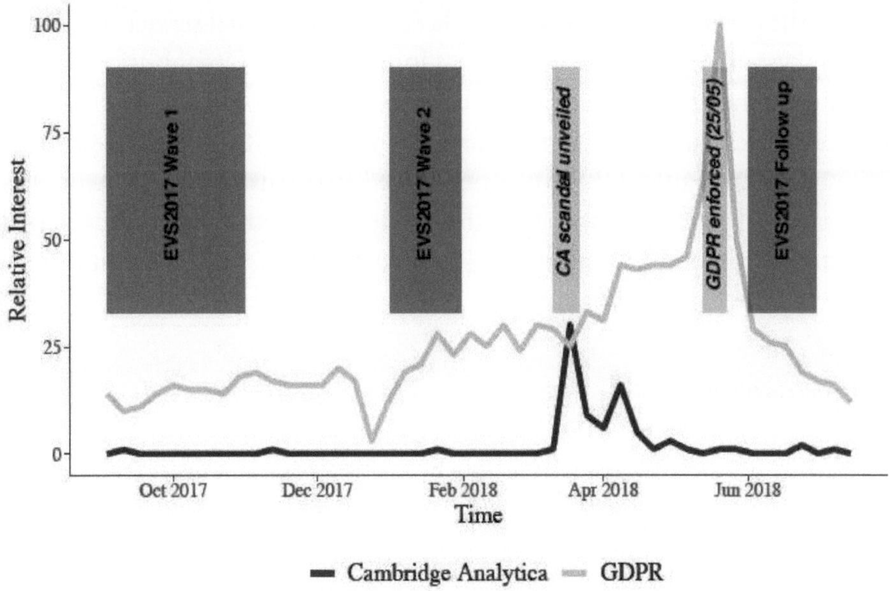

FIGURE 2.1 Timeline of the study (source of Google search volumes: Google Trends).

The interest-over-time chart should be interpreted as follows: 'Numbers represent search interest relative to the highest point on the chart for the given region and time. A value of 100 is the peak popularity for the term. A value of 50 means that the term is half as popular. A score of 0 means there was not enough data for this term.' Google also provides additional search queries that looked for Cambridge Analytica and GDPR. In the case of Cambridge Analytica, in 100 percent of the search terms, the term 'Facebook' was included; in the case of GDPR, 48 percent of the search were accompanied with 'privacy'. Both cases indicate the relevance for this study.

traditional types of institutions, we consider the other trust items. Trust in the following institutions is measured: the church, the armed force, the education system, the press, trade unions, the police, parliament, civil service, the social security system, the European Union, United Nations Organization, health care system, the justice system, major companies, environmental organizations, political parties, and government. Correlations between trust items are visually represented in Figure 2.2. It can be noticed that trust in social media has relatively low correlations with the other institutional trust items. The internal consistency of the scale is quite high (Cronbach's alpha = 0.88), but it may be inflated by the high number of items considered.

The Correlates of Trust

Social media use is measured by asking how often respondents follow politics via social media. Response categories range between 1 (never) and 5 (every day). Although this indicator does not fully capture the extent of social media usage, this is the best proxy available in the questionnaire.

To test the trust-nexus hypothesis, we are interested in the extent to which institutional trust well as social trust are related to trust in social media. For institutional trust we use a mean score of trust in institutions that are not social media, depending on the results of a factor analysis (see below)[10]. Social trust is measured with the question 'Generally speaking, would you say that most people can be trusted or that you can't be too careful in dealing with people' and two answer categories: 0 (can't be too careful) and 1 (most people can be trusted).

[10] Alternative models including trust in government as a single variable instead of the composite measurement of institutional trust are presented in Appendix A as they yield different results in the last part of the analyses. Trust in government and institutional trust are strongly correlated ($\rho = 0.76$)

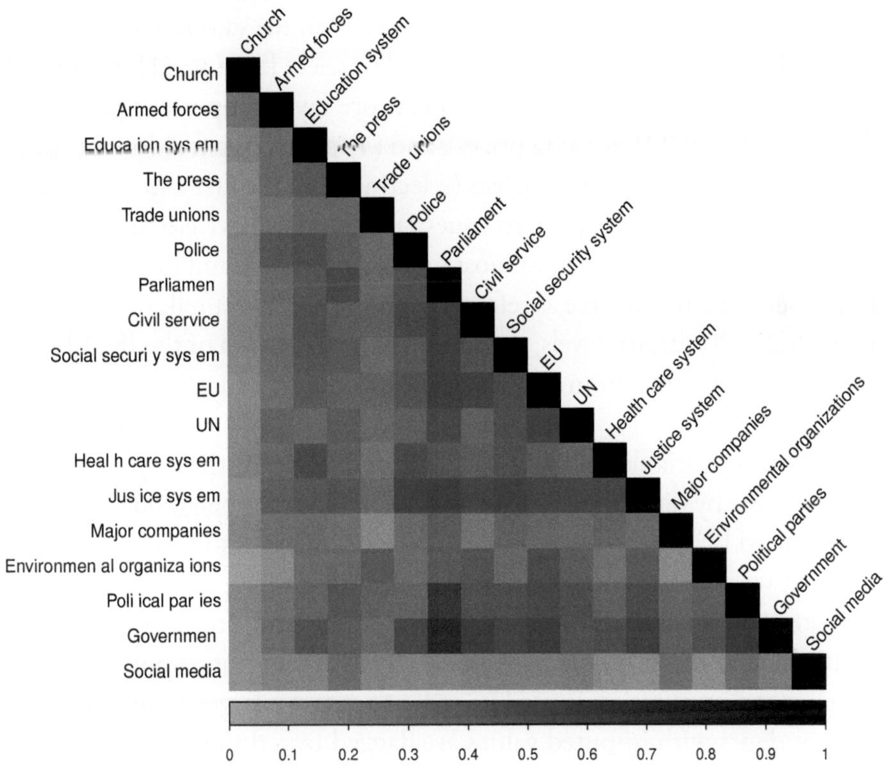

FIGURE 2.2 Correlation plot among institutional trust items.

As indicated by the legend, the lighter color, the weaker the correlation (Pearson's rho), and viceversa: the darker the color, the stronger the correlation (N = 1,361).

As for the cultural explanation of post-materialism, we include the traditional post-materialism index, following Inglehart's two-item index. The question is 'If you had to choose, which one of the things on this card would you say is most important?' followed by a question on the second most important issue. The four issues respondents needed to select from are (1) 'maintaining order in the nation', (2) 'giving people more say in important government decisions', (3) 'fighting rising prices', and (4)'protecting freedom of speech'. We distinguish between materialists (selecting both the first and third issues as most and second most important issues), postmaterialists (selecting the second and fourth as most and second most important issues), and those with mixed responses (which will serve as reference category). To test reflexive modernity, we include educational levels, coded using an adaptation of the ISCED 2011-classification, ranging from 0 (not completed primary education) to 7 (master's and above).

As control variables, we consider age, derived from the respondent's age of birth; sex, by distinguishing men (code 0) from women (code 1). Income is measured by the question 'Here is a list of incomes and we would like to know in what group your household is, counting all wages, salaries, pensions and other incomes that come in.' Answer codes range from 1 (lowest decile) to 10 (highest decile). Due to the considerable item-nonresponse (13.9 percent), mean imputation is used in order to retain a larger analytical sample. Respondents with imputed values are flagged by a dummy included in the models, which allows to rule them out when assessing the impact of income on trust in social media. For work status we distinguish between those employed (reference category) and unemployed, students, retired, and those in another category. We also control for the frequency of following politics on tv, as a general measure of media exposure[11]. Descriptive statistics of all variables can be found in Tables A.2 and A.3 (see Appendix A).

[11] Controlling for political news consumption via television might also imply that we have

Analytical Strategy

Because we gradually build up insights into endogenous and exogenous explanations of trust in social media, the analytical section is incremental. The first step involves a principal component factor analysis on the trust-items. A model with one factor is tested and evaluated on the basis of a) the correlations of the factors with each item (represented by the factor loadings) and b) the unexplained variance of each item (uniqueness). Additionally, an alternative model retaining all the factors with an eigenvalue > 1 and using Varimax-rotation (to allow for maximum variation between factors) is assessed. Concerning the debate over the dimensionality of trust in institution (i.e. whether trust in social media reflects general opinions about institutional quality, or rather is a rational consideration about the functioning of each institution), this procedure should highlight the extent to which trust in social media is distinct from trust in other institutions. This analysis will be done on the pre-test data, using pairwise missing deletion[12] (N = 1,504).

In a second step, we examine individual variation in trust in social media. Since our dependent variable has four ordered response categories, we perform an ordered logistic multiple regression analysis, employing the pre-test data (N = 1,504).[13]

In the final part of the analysis, we tackle the question whether respondents have become more skeptical towards social media after the data privacy contro-

a better estimate of social media use, as it might parcel out news consumption of our independent variable 'following politics on social media'

[12] The procedure suggested in https://stats.idre.ucla.edu/stata/faq/how-can-i-do-factor-analysis-with-missing-data-in-stata/ has been adopted.

[13] The proportional odds assumption was tested using the function 'gologit2, autofit' from the homonymous Stata package, and the insignificant test statistic indicated that none of the main independent variables violates the parallel lines assumption (χ^2=17.54, df=12, $p = 0.13$).

versy in the spring of 2018. To do so, we use the panel structure of the Dutch part of the EVS 2017: the responses of the respondents questioned in June 2018 (post-test) are matched to their responses in the pre-test. After displaying individual change using cross-tabulation, the propensity to display changes in trust is explored. Because of the limited four-categories response scale, in this exploratory step we perform a change scores analysis (van Ingen & Bekkers, 2015), by calculating whether respondents have become more trusting (higher score in post-test than in pre-test), more distrusting (lower score in post-test than in pre-test) or unchanged (same levels of trust in post-test and in pre-test). To test who has become more trusting or more distrusting over time, multinomial regression analysis is performed, with relevant independent variables from the cross-sectional analysis added as explanations. The merged pre- and post-test datasets are used (N= 1,097).[14]

For the analyses, we use Stata, version 16. Additionally, Figure 2.1 and Figure 2.2 are produced via RStudio.[15]

4 • RESULTS

Is Trust in Social Media Different from Trust in Other Institutions?

Table 2.1 shows the result of two principal component factor analyses. In the 1-factor model, the unidimensionality question is directly tackled. Results show that all items but trust in social media and the church may be considered

[14] Despite the large drop in the number of cases between the pre- and the post-test due to attrition and item non-response, both a t-test ($t = 1.32$, df=1,502, Pr($|T| > |t|$)=0.18) and a chi-square test (χ^2=3.82, $p = 0.281$) indicate no substantial differences in the level of trust in social media of the dropped 407 cases compared to the 1,097 respondents constituting the analytical sample of the post-test.

[15] The R package gtrendsR (Massicotte & Eddelbuettel, 2021) was used to download the Google Trends data and produce Figure 2.1.

as belonging to an unidimensional 'institutional trust' factor, explaining 35% of the variance. For trust in the church and trust in social media, the low factor loadings (λ = 0.22 and λ = 0.26, respectively), and the high uniqueness (0.95 and 0.93, respectively) suggest that they do not belong to the scale.

An alternative model, retaining three factors displaying an eigenvalue >1 was also estimated. Our primary interest lies in the item on trust in social media – the last item offered in the rating scale. Results show that the third factor loads rather strongly on trust in social media (λ = 0.84) and has moderate correlations with trust in major companies (λ = 0.46), political parties (λ = 0.41), and, to a lesser extent, trust in the press (λ = 0.38). These latter three items, however, display cross-factor loadings, as they are also relatively strongly correlated with the other factors, thus making the substantive interpretation of this factor rather complex.

The first factor captures the largest portion of variation, and shows loading on several items, most notably (ranked from strongest to weakest loadings) trust in government and in the parliament (λ = 0.72), the European Union, parliament, environmental organizations, civil service, political parties, the justice system, United Nations organizations, social security system, trade unions, the police, the press, the education system and the health care system (λ = 0.40). The latter also has a stronger loading (λ = 0.58) on the second factor. The first factor could be referred to as 'institutional trust' and is used as such in subsequent analyses; it should be noted, however, that it is distinct from the partisan institutions described by Rothstein and Stolle (2008) because of the presence of environmental organization and social security system. The second factor relates to order and 'neutral' institutions (in line with Rothstein & Stolle, 2008), as it combines trust in the armed forces, health care system, police, church, education system and justice system.

Altogether, the result of the factor analysis demonstrated that trust in social media can be treated as distinct from other types of trust in institutions. This first piece of evidence does not support trust in social media as being part of

TABLE 2.1 Factor loadings and uniqueness after principal components factor analysis (N = 1,504).

Trust in...	1-factor solution		3-factor solution			
	Factor	Uniqueness	Factor 1	Factor 2	Factor 3	Uniqueness
Church	0.22	0.95	−0.08	**0.52**	0.21	0.68
Armed Forces	**0.43**	0.81	0.11	**0.67**	0.13	0.52
Education System	**0.57**	0.67	**0.42**	**0.44**	0.03	0.62
The Press	**0.54**	0.70	**0.43**	0.19	0.38	0.64
Trade Unions	**0.45**	0.80	**0.49**	0.00	0.11	0.75
Police	**0.61**	0.62	**0.43**	**0.56**	−0.07	0.49
Parliament	**0.78**	0.40	**0.72**	0.25	0.19	0.38
Civil Service	**0.68**	0.54	**0.67**	0.21	0.03	0.50
Social Security System	**0.65**	0.58	**0.56**	0.31	0.11	0.57
European Union	**0.70**	0.52	**0.71**	0.14	0.12	0.47
United Nations	**0.64**	0.59	**0.56**	0.24	0.20	0.58
Health Care System	**0.60**	0.64	**0.40**	**0.58**	−0.05	0.50
Justice System	**0.71**	0.49	**0.63**	**0.42**	−0.06	0.43
Major Companies	**0.44**	0.81	0.16	**0.41**	**0.46**	0.59
Environmental Org.	**0.52**	0.73	**0.70**	−0.15	−0.05	0.49
Political Parties	**0.69**	0.53	**0.63**	0.11	**0.41**	0.42
Government	**0.76**	0.42	**0.72**	0.24	0.14	0.40
Social Media	0.26	0.93	0.08	−0.01	**0.84**	0.29

In bold: factor loadings $\geq$ 0.4; Source: EVS 2017 Netherlands

an abstract idea about institutional quality; rather, it might be a reflection of perceptions of its functioning.

Who Trusts Social Media?

Before highlighting the covariates of trust in social media, it is important to look at the distribution of this variable. A univariate exploration of the pre-test data reveals that trust in social media generally is low: approximately 22.8

percent of the respondents has no trust at all in social media, 64.2 percent has not very much, 12.3 percent quite a lot, and 0.7 percent a great deal of trust in social media. In fact, as Appendix Table A.1 shows, of all listed institutions, none is ranked as low as social media and the church: on a scale from 1–4, the most trusted institution is the health care system (2.870; sd = 0.65), while social media ranks at the bottom (1.91; sd = 0.60) together with trust in the church (1.91, sd = 0.82). Thus, ahead of the data breach controversies, the Dutch were already quite wary of social media.

Turning to theoretically relevant explanations[16], in a first model (see Table 2.2), we test the impact of the use of social media. In line with the expectations (see Hypothesis 2), those who frequently follow politics via social media have more trust.[17] It is worth noticing that receiving political news via the more traditional television outlets correlates significantly with lower trust in social media. While causality issues might be at play (as individuals distrusting social media might turn to traditional media, or at least turn away from social media, for their political news), the evidence nonetheless suggest a tense relationship between trust and media usage.

Secondly, we expected that trust in social media would be part of a trust-nexus: trust in social media would flow from other types of vertical (institutional) and horizontal (social) trust. The analysis, presented in Model 2 of

[16] Regarding the control variables, no gender differences in trust in social media are reported. Age and income are both unrelated to trust in social media. However, for work status, there are significant effects worth reporting: both the unemployed and the retired display elevated levels of trust in social media. The interpretation is a bit difficult: on the one hand, being retired is a relevant employment status, on the other hand, it combines the eldest respondents into one category, meaning that this effect should also be as if the elderly have more trust in social media than the rest of the population.

[17] This finding should be interpreted in terms of correlation, as it is not possible in the current research design to disentangle the exact direction of the causal relationship: in other words, it is possible that those who display more trust in social media are more prone to use social media to follow politics.

Table 2.2, provided mixed results. In line with Hypothesis 3, we discerned a positive and strong correlation of institutional with trust in social media[18]. However, social trust did not significantly correlate with trust in social media, implying that those generally trusting people have the same levels of trust in social media compared to those not trusting other people. We thus reject Hypothesis 4, since there is no spill-over-effect of social trust on trust in social media.

The role of cultural indicators is tested in Model 3. We found that the correlation between trust in social media and Inglehart's postmaterialization-index, synthesized in Hypothesis 5, is confirmed: compared to those with a mixed value pattern, materialists are more likely to trust social media. Postmaterialists, however, are not significantly different from those with a mixed response type. Testing the reflexive modernity thesis, the level of education shows a negative and significant effect, indicating that the higher educated are less trusting of social media than the lower educated, in line with the expectation (see Hypothesis 6). The full model (see model 4 in Table 2.2) confirms the results of the previous models.

Does Trust in Social Media Change Over Time?

Lastly, we exploit the panel structure of the data. As we can see from Table 2.3, some individual change took place over time: About 27 percent of the respondents became more trusting in social media (below diagonal, marked in light grey), while about 22 percent turned more distrustful. Approximately 51 percent of the population did not change opinion over time (main diagonal, marked in white). The cross-tabulation provides first evidence against hypothesis 7, or at least suggests nuance: people have not *en masse* become

[18] Similar results are achieved when using only trust in government, see Table ?? in Appendix A.

TABLE 2.2 Ordinal logistic regression of trust in social media on relevant covariates.

Independent variable	Model 1: Use	Model 2: Trust-nexus	Model 3: Cultural	Model 4: Full model
Follow politics on social media[a]	0.320***			0.348***
	(0.045)			(0.045)
Institutional trust[a]		1.285***		1.513***
		(0.135)		(0.144)
Social trust[a]		−0.178		−0.100
		(0.124)		(0.126)
Postmaterialism (Ref: Mixed)				
- Materialism			0.447**	0.489**
			(0.150)	(0.152)
- Post-materialism			0.042	0.018
			(0.147)	(0.151)
Levels of education[a]			−0.104***	−0.188***
			(0.031)	(0.033)
Age[a]	0.004	0.001	−0.005	0.001
	(0.005)	(0.005)	(0.005)	(0.006)
Woman (Ref = Man)	−0.011	−0.047	−0.014	−0.056
	(0.111)	(0.112)	(0.111)	(0.113)
Income[a]	0.005	−0.030	0.026	0.023
	(0.022)	(0.022)	(0.023)	(0.024)
Income missing dummy	0.018	0.165	0.019	0.023
	(0.156)	(0.156)	(0.156)	(0.159)

Table continued on next page

Work status (Ref: Employed)				
- Unemployed	0.829**	0.818**	0.908**	0.699*
	(0.310)	(0.310)	(0.312)	(0.316)
- Student	0.031	−0.322	0.056	−0.356
	(0.314)	(0.320)	(0.316)	(0.326)
- Retired	0.492**	0.482**	0.459*	0.462**
	(0.184)	(0.185)	(0.183)	(0.187)
- Other work status	0.355	0.513**	0.321	0.326
	(0.197)	(0.197)	(0.196)	(0.201)
Follow politics on television[a]	−0.165***	−0.138**	−0.014	−0.200***
	(0.046)	(0.044)	(0.045)	(0.048)
Cut-off trust in social media = 1	−1.118***	−1.131***	−0.972***	−1.166***
	(0.165)	(0.166)	(0.168)	(0.172)
Cut-off trust in social media = 2	2.165***	2.217***	2.253***	2.388***
	(0.175)	(0.176)	(0.179)	(0.184)
Cut-off trust in social media = 3	5.534***	5.616***	5.611***	5.852***
	(0.388)	(0.390)	(0.391)	(0.394)
Observations	1.504	1.504	1.504	1.504

Entries represent ordered log-odds regression coefficients (standard errors in parentheses). Analysis controlled for matrix design group. Source: EVS 2017 Netherlands.
[a] Variable centered around the mean | $^{*}p < 0.05$; $^{**}p < 0.01$; $^{***}p < 0.001$

more distrusting over the data privacy controversy; rather, a majority did not adjust their trust in social media.

Despite the loss of information, combining respondents into three coarser groups (increased trust, no change, decreased trust) seemed appropriate given the low number of respondents in some of the cells (see Table 2.3).

Table 2.4 presents the results of an exploratory multinomial logit regression model to analyze who changed trust between the pre- and the post-test. The Nagelkerke pseudo R-squared indicates that the model explained approximately 9 percent of the likelihood of change in trust in social media. Looking

TABLE 2.3 Cross Tabulation of Trust in Social Media in the pre-test and in the post-test.

Trust in Social Media post-test	Trust in Social Media pre-test				
	Not at all	Not very much	Quite a lot	A great deal	Total
Not at all	52 20.80%	122 17.10%	22 16.90%	0 0.00%	196 17.90%
Not very much	172 68.50%	493 69.20%	93 72.50%	4 100.00%	761 69.40%
Quite a lot	26 10.40%	96 13.50%	15 11.50%	0 0%	137 12.50%
A great deal	1 0.40%	1 0.10%	0 0%	0 0%	2 0.20%
Total	250 100.00%	712 100%	130 100%	4 100%	1,097

Source: EVS 2017 Netherlands.

at explanations for increased trust over time, a few interesting patterns emerged. First and foremost, we found that higher institutional trust is correlated with a higher likelihood of decreased trust in social media and, viceversa, a lower likelihood to display increased trust in social media.[19] The impact of institutional trust on the change in trust in social media is also presented in Figure 2.3. It appears like a strong institutional trust, instead of buffering the impact of a breach of trust, makes people more skeptical of social media, perhaps because of the stronger disappointment arising from their higher expectations on the functioning of institutions.

Results also show a positive correlation between following political news on

[19] Coefficients result, however, not significant when a measure of trust in government alone is used, see Table A.5 in Appendix A.

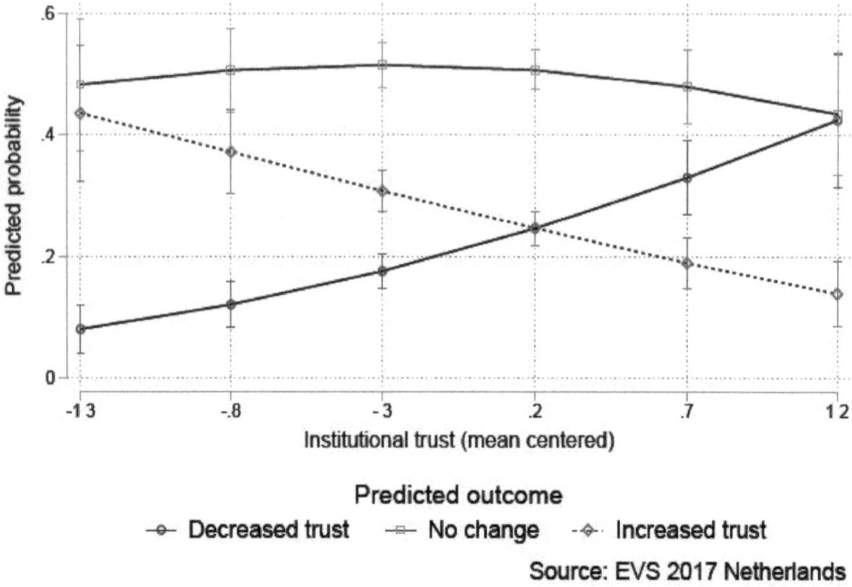

FIGURE 2.3 Predicted probabilities of change in trust in social media and 95% confidence intervals by institutional trust (estimated from the model in Table 2.4).

tv and having more trust in social media over time; the likelihood of having more trust over time, compared to no trust adjustments, is also lower for unemployed, retired people (thus mostly elderly) and people with another work status. As for to the likelihood of adjusting trust downwards over the data privacy controversy, the following is found: firstly, a more frequent use of social media is positively associated with the chance of decreased trust. Thus, although the use of social media resulted positively correlated with trust in social media in the pre-test (see Table 2.2), those who used social media more frequently were also more inclined to become distrustful of social media over the privacy controversy. Secondly, a materialist value pattern is associated with decreased trust; although in the pre-test materialists appeared more trusting (see Table 2.2), the post-test showed that they are more likely to adjust their trust downwards during the privacy controversy.

TABLE 2.4 Multinomial logistic regression of change in trust in social media (Reference = No Change).

Independent variable	Decreased trust	Increased trust
Follow politics on social media[a]	0.132*	−0.092
	(0.064)	(0.062)
Institutional trust[a]	0.733***	−0.440*
	(0.210)	(0.193)
Social trust[a]	−0.035	−0.063
	(0.185)	(0.171)
Materalism index (Ref: Mixed)		
- Materialist	0.510*	0.362
	(0.214)	(0.207)
- Postmaterialist	0.157	0.162
	(0.221)	(0.207)
Educational level[a]	−0.041	0.081
	(0.047)	(0.045)
Age[a]	0.002	0.000
	(0.008)	(0.008)
Female	−0.291	−0.023
	(0.167)	(0.155)
Income[a]	−0.019	−0.021
	(0.035)	(0.033)
Income missing	0.176	0.499*
	(0.235)	(0.209)

Table continues on next page.

Work status (Ref: Employed)		
- Unemployed	−0.087	−1.129*
	(0.418)	(0.521)
- Student	−1.279*	−0.419
	(0.600)	(0.447)
- Retired	−0.099	−0.528*
	(0.272)	(0.252)
- Other work status	−0.349	−0.709*
	(0.299)	(0.286)
Follow politics on television[a]	−0.076	0.226***
	(0.072)	(0.067)
Intercept	−0.980***	−0.555*
	(0.255)	(0.229)
Nagelkerke's R2	0.094	
Observations	1,097	

Entries represent ordered log-odds regression coefficients (standard errors in parentheses). Analysis controlled for matrix design. Source: EVS 2017 Netherlands.
[a] Variable centered around the mean | * $p < 0.05$; ** $p < 0.01$; *** $p < 0.001$

5 • CONCLUSION AND DISCUSSION

Citizens of the world are increasingly connected by the World Wide Web for social interactions and for information gathering. As people increasingly rely on online social media, the question also arises whether netizens actually trust these novel institutions. The literature on institutional trust has shown that trust rarely is an imprint of the functioning of the evaluated institutions; rather, it is an expression of a myriad of exogenous factors. In 2018, events shedding light on the flaws of protection of online data privacy – an element expected to be at the core of trust in social media – led to ask whether an increased salience of user data privacy would result in more distrust in social media. Extending the Dutch data collection of the EVS 2017 with a panel structure allowed for a rather nuanced response.

First, among a sample representative for the Dutch population, trust in social media is rather low; social media even rank at the bottom of all institutions administered in the survey. Further, trust in social media is not strongly interlinked with trust in other types of institution, weighing in in ongoing debates about the dimensionality of trust: if responses on trust in institutions would reflect general evaluations of institutional quality, social media are clearly not part of this evaluation. Combined, this would provide evidence for endogenous explanations, namely that people have a distinct and rational assessment of social media. Nevertheless, the strongest argument in favor of an endogenous explanation is not supported by our analysis, as people have not overwhelmingly become more distrusting towards social media in response to the turmoil over online data privacy. Half of the respondents showed no change in trust in social media, while the remaining part was roughly equally split over decreased as well as increased trust. It should be considered that over the period under consideration, the Dutch were not only confronted with the negative side of the data privacy controversy, yet also experienced the introduction of protective GDPR-regulation and the actions undertaken by Facebook and other social media to strengthen the control over personal data. Also, in the aftermath of the Cambridge Analytica scandal, Facebook apologized (cf. Hall, 2020), and an apology could, under some circumstances, restore trust (cf. Ayaburi & Treku, 2020). Nevertheless, despite the increased salience of online data privacy, we did not witness a significant drop in trust in social media.

Before disqualifying endogenous explanations altogheter, we suspect the existence of a floor effect, whereby trust in social media is already at a low level and cannot fall below this threshold. We found that those who have high institutional trust, follow politics via social media more frequently, and people with a materialistic value pattern, were more likely to adjust their trust downwards over the data privacy controversy, compared to maintaining a stable level of trust. Notably, these were also the people who displayed higher trust in social media in the pre-test. This suggests that social groups may adjust their trust evaluations in reaction to privacy threats in different ways, something that should be explored further. Along these lines, the hypothesis that people react differently to privacy threats depending on their level of education will be examined in Chapters 4 and 5. Finally, the relationship between use of social media and trust may provide fertile ground for future research: abstaining from any causal claim, we found an association between using and trusting social media, as those who follow politics via social media had more trust, whereas television news consumers displayed less trust in social media.

The individual-level correlates support the existence of exogenous sources of trust in social media. First of all, we found evidence of a vertical trust-nexus: if people have positive attitudes towards political and civic institutions, they also have higher trust in social media. Second, cultural roots underlie trust in social media, as materialists as well as the lower educated report higher trust levels. It appears like trust in social media has a myriad of causes, with perceptions of its functioning being only a minor factor.

Our main findings, that trust in social media is low, and that the turmoil over online data privacy did not cause a massive drop in trust in social media, presents us with some interesting challenges. First, besides the low trust in social media, people are still massively connected using Facebook, Twitter, Instagram, and the like. This would be in line with the so-called 'privacy paradox', according to which there is a discrepancy between privacy concerns and actual privacy behaviors (Kokolakis, 2017): along these lines, perceiving privacy risks related to social media may not necessary lead to adjust behaviors, e.g. by stopping using social media, nor translate into a negative cognitive evaluation of social media's performance. As pointed out by Brown (2020), the stakes of leaving Facebook are very high due to the loss of social connections, and may hinder drastic adjustments. Second, there may be something distinctive in the Dutch context which affects our findings, since the Dutch society is considered to be a high-trust, 'open curtains' society' (cf. Mols & Janssen, 2017): for instance, the erosion of political trust occurred in the Netherlands in a delayed fashion compared to the other Western countries (Bovens & Wille, 2008). Accordingly, the Dutch may be less sensitive to short-term endogenous adjustments of trust in social media compared to the expectations.

For future research we would primarily suggest to improve the measurement of trust in social media. Although a study fielded within CRONOS showed that people first and foremost consider Facebook when confronted with this item, there are many social media respondents may have in mind when answering; in the context of our study, this is also a limitation considering that the CA scandal involved specifically Facebook. Moreover, in addition to trust-measures, behavioral items like effective internet use might be useful. Lastly, qualitative data collected among those who closed their Facebook-account in the aftermath of the data privacy controversy might unveil different mechanisms, which may be otherwise difficult to capture with a survey.

Public acceptance of a COVID-19 Health Pass

Evidence from a vignette study in the Netherlands

ABSTRACT

In attempt to safely resume social life amidst the COVID-19 pandemic, governments adopted Health Passes, e.g. digital certifications of low risk of carrying a coronavirus infection as a result of vaccination, negative test result or recent healed infection. Yet, the deployment of these tools is accompanied by concerns over the privacy risks they entail. In this study, we evaluate the acceptability of a COVID-19 Health Pass in the Netherlands by means of a vignette experiment administered in May 2021 to a representative sample of the Dutch population via the LISS panel. The acceptability of the measure is investigated as a function of (a) the informational norms associated with its features according to the contextual integrity framework; (b) the individuals' predispositions towards the institutions promoting it. Results show that the measure is largely supported by Dutch citizens and that the support is not eroded by more privacy-intrusive features of the Pass itself. Additionally, institutional trust fosters the acceptability of the technology. Findings are discussed in light of the adopted theoretical framework, and of the emergency context in which the vignette experiment took place.

This chapter is a joint work with Tim Reeskens. Replication materials for this chapter are available on OSF (https://edu.nl/mba7a).

1 • INTRODUCTION

The deployment of digital health surveillance technologies has been central in the fight against the COVID-19 pandemic (Calvo et al., 2020; Lewandowsky et al., 2021; Newlands et al., 2020). To allow people to resume social activities and international travels, governments have introduced COVID-19 health passes (hereafter: HP), i.e. a certification, often in a digital form, of an individual's low risk of infection (e.g. after a negative COVID-19 test result, a recent coronavirus infection or the vaccination[1]). Intended to promote public health, increased freedom, and economic benefits, the COVID-19 HP has been criticized, among other reasons because it raises privacy concerns (Ada Lovelace Institute, 2021; Newlands et al., 2020). Experts claim that the COVID-19 HP carries high potential of privacy invasion due to the sensitivity of the information exchanged, the data-exhaust it generates (see also the discussion on dataveillance in Chapter 1) and the wide array of potential actors accessing the information. For instance, to access social venues like museums or concerts, the identity check of the pass holder is delegated to third parties (e.g. waiters or bouncers) not always equipped to properly handle sensitive personal data. Finally, some have warned that the rushed implementation of these novel technologies without a refined regulatory framework may generate a 'surveillance creep', i.e. the risk that an intrusive surveillance policy enforced due to an emergency remains in place after the emergency is passed (Calvo et al., 2020; Vitak & Zimmer, 2020).

Because of the novelty of the COVID-19 HP, widespread insights into its public acceptability are scarce, even though its introduction may not be effective unless the measure is endorsed and actively used by large segments of

[1] It should be mentioned that the vaccination passport is not new, as vaccinations are usually registered on a paper support (e.g. in the Netherlands, the 'geel vaccinatieboekje') to be showed in some, limited circumstances.

the population.[2] Therefore, the aim of this chapter is to contribute to better understand under what conditions the populace will endorse the COVID-19 HP with regards to its potential for privacy violation. The rapid and controversial introduction of such tools offers a unique opportunity to delve into the acceptance of (digital) surveillance among the population and investigate the extent to which people value their privacy in emergency situations. We argue that, in line with the Contextual Integrity framework (Nissenbaum, 2010, 2011, 2019), the acceptability of the COVID-19 HP depends on informational norms regulating the exchange of information enabled by the technology. In addition, we complement this perspective by investigating how such norms are filtered by the perception of risk associated with the data exchange and individual cultural predispositions – in particular, trust in institutions. Similarly to Chapter 2, we focus on the Netherlands, a country in which the introduction of a digital health surveillance tool is facilitated by high levels of digitalization among individuals, firms, and governmental institutions.[3] The trends in the spread of COVID-19 in the country have followed those of most European countries, and there has been strong opposition among specific segments of the population to the more invasive containment measures[4] such as the curfew, whose introduction on January 23rd, 2021 caused riots in some Dutch cities.[5]

To answer our research questions, we implemented a vignette experiment in which we describe a COVID-19 HP and varied some of its features based on a theoretical framework. The experiment was administered to a probabilistic

[2] Parallel discussions have taken place for the coronavirus contact tracing apps (cf. Altmann et al., 2020; Buder et al., 2020)

[3] See https://ec.europa.eu/digital-single-market/en/scoreboard/netherlands.

[4] See an overview on https://graphics.reuters.com/world-coronavirus-tracker-and-maps/countries-and-territories/netherlands/.

[5] See, e.g., https://www.theguardian.com/world/2021/jan/26/netherlands-third-night-riots-covid-curfew-lockdown-protesters.

sample of individuals residing in the Netherlands via the LISS Panel[6] in May 2021, when information on the COVID-19 HP and its nationwide implementation was still scarce.[7] Fielding such an experiment in the LISS Panel enables to also consider individual characteristics which may affect the acceptance of the COVID-19 HP. Ultimately, the experiment contributes to gain insights into the way individuals form privacy evaluations in information-intense societies, which could also explain the marginalization of specific groups as societies find ways to resume social activities amidst a pandemic.

In the remainder, we start by outlining the theoretical framework explaining public acceptance of the COVID-19 HP, focusing on the role of privacy evaluations and on cultural predispositions towards institutions enabling health surveillance. In the subsequent section, we translate the theoretical insights into specific informational norms guiding the acceptability of the COVID-19 HP, and formulate hypotheses. The vignette experiment is then described alongside the other measurements and analytical strategy. After presenting the results, the chapter closes with a summary of findings and discussion.

2 • ACCEPTING THE COVID-19 HEALTH PASS

Privacy trade-offs and Contextual integrity framework

The conditions under which people accept to disclose private information are a central object of study in contemporary information-intense societies. The privacy-security trade-off approach, first and foremost, has been applied to understand the acceptance of surveillance: individuals are willing to renounce

[6] More information about the LISS panel can be found at: www.lissdata.nl.

[7] In July 2021, the HP has been introduced in the Netherlands via the CoronaCheck app for international travels and some events. From September 25th, 2021, up to February 25th, 2022, the HP was mandatory upon entry in restaurants, cinemas, theatres, etc.

to some of their privacy in exchange for more security (Pavone & Degli Esposti, 2012), especially in situations of uncertainty or emergency. Along these lines, Davis and Silver (2004) found that in the aftermath of 9/11 the sense of threat and the willingness to renounce civil liberties went hand in hand among US citizens. The implication for the introduction of the COVID-19 HP is that individuals are willing to accept any exchange of personal information insofar as that allows them to reduce the risks of infection by the novel coronavirus while resuming social activities. Polish findings indeed suggest that emergency-related feelings, such as perceived threat and lack of control, were positively related to the support of privacy-invasive surveillance technologies aimed at fighting the pandemic as they are seen as ways to reduce uncertainty (Wnuk et al., 2020).

Unlike the trade-off perspective, which often relies on abstract perceptions of security and privacy (Pavone & Degli Esposti, 2012), the privacy calculus perspective sees privacy decisions as dependent upon a rational calculus of costs and benefits associated with the disclosure of information, considering the specific circumstances in which the decision occurs (Hallam & Zanella, 2017). Marwick and Hargittai (2018) untangled that US students made cost-benefit evaluations when deciding whether to disclose personal information online by considering the type of information and the institutional actors involved. Further, even when confronted with an actual privacy breach such as the Cambridge Analytica scandal, social media users have been found to keep their accounts to avoid the loss of social interactions (Brown, 2020).

The framework we rely on in this study is the one of contextual integrity (CI) (Nissenbaum, 2004, 2010, 2011). This framework systematizes the elements that individuals consider when evaluating whether a new technology is acceptable in terms of disclosing personal data to serve its purpose. The argument is that privacy considerations depend upon informational norms, which are embedded in context-specific (social) norms, roles, values and purposes (Nissenbaum, 2010, 2019). In this perspective, privacy violations are seen as violation of informational norms rather than as general disclosure of personal information!(Nissenbaum, 2010).

The concept of informational norms implies that, depending on the specific circumstances, individuals hold expectations over the appropriateness of the transmission and distribution of personal data. Accordingly, a new privacy-intrusive system or technology will be accepted by the potential user if it complies with the informational norms of the context in which the technology is used (i.e., the contextual integrity); the same technology in a different context may be received differently. Existing research has shown the relevance of the CI framework in the public acceptance of recent technological innovations, such as smart home devices (Apthorpe et al., 2018; Horne et al., 2015), health-based applications (Gerdon et al., 2021; Y. J. Park & Shin, 2020), and COVID-19 contact tracing apps (Vitak & Zimmer, 2020).

Following the CI framework, within each specific context informational norms are a function of five key parameters (Nissenbaum, 2019). First, informational norms depend on the three actors involved, including the subject whom the information relates to, the recipient of the information and the sender who transmits the information. Second, informational norms are a function of the attributes, namely type/content of information transmitted. Finally, they depend on the transmission principle, e.g. the strategy of dissemination or the retention period of the personal information. To exemplify, one may find appropriate to share their telephone number (attribute) with a new acquaintance (recipient) to exchange informal communications (purpose) via one-to-one interactions (transmission principle). Tweaking one parameter can result in a different decision: for instance, if the recipient changes to one's employer, the individual may not be willing to share the same information any longer. Accordingly, the recipient, attributes, and transmission principles linked to the information exchange enabled by the COVID-19 HP, alongside the purpose of its introduction, are expected to affect the public acceptance of the COVID-19 HP, which is the object of the present study.

While the CI framework explains how the features of a technology itself affect its acceptability, it is important to also consider the cultural predispositions held by the individuals adopting the technology, as well as their risk perceptions. As extensively explained in Chapter 1, datafication-induced risks like the ones involved in the COVID-19 HP data exchange tend to be invisible and require an exercise of reflexivity to be acknowledged. On the one hand, the emergency situation of the pandemic may contribute to the latency of datafied risks (cf. Chapter 1). On the other hand, some individuals may be more prone to acknowledge these risks due to their predispositions, affecting not only the overall acceptability of privacy-intrusive technologies, but also the perception of the informational norms associated with them. In the case of the nationwide introduction of the COVID-19 HP, which creates uncertainties for the subjects by enhancing the control capabilities of the institutions surveying (Trüdinger & Steckermeier, 2017), we foremost need to consider individual orientations towards these institutions.

Institutional trust

Institutional trust has been found to work as a decisional heuristic in situations of uncertainty (Bradford et al., 2020; Trüdinger & Steckermeier, 2017). According to Mayer et al. (1995), evaluations of trustworthiness consist of three elements (Hendriks et al., 2016): ability (i.e., to effectively perform the task), benevolence (i.e., the institution does not want to harm the trustor) and integrity (i.e. compliance with shared norms and principles). Accordingly, when the involved institutions are deemed trustworthy, the appropriateness of the COVID-19 HP is enhanced by the perception of expertise with handling the data, good intentions in setting up the surveillance measure, and normative alignment, i.e. the belief that the institution under scrutiny complies with societal expectation (Bradford et al., 2020) and will not abuse the exchanged data.

In particular, there are two institutions whose trustworthiness is relevant when investigating the public acceptability of a COVID-19 HP: the government and science. Whilst the government is the institution primarily in charge of the deployment of the measure, scientific experts have been at the forefront throughout the pandemic, providing recommendations to the governments and playing an active role in informing the public. Though there is a debate in the literature over whether individuals differentiate among institutions when evaluating their trustworthiness (see, e.g., Fisher et al., 2010; Hooghe, 2011) and what the distinction is based on (Newton et al., 2018; Rothstein & Stolle, 2008), we adopt an exploratory approach and look at the two institutions separately. While the government is a partisan institution and its assessment may be dependent upon political evaluations and subject to short-term variations over time (Reeskens et al., 2021), science is an impartial institution whose trustworthiness is rooted in more general, 'default' evaluations (Hendriks et al., 2016, p. 151).[8]

To sum up, this study aims at understanding the public acceptance of the COVID-19 HP by complementing the CI framework, which focuses on the way the features of the technology itself affect the perception of privacy violation risks, with an individual-level perspective which accounts for the subjective predispositions towards the institutions promoting the measure, i.e. the government and science.

3 • THE PRESENT STUDY

To test the influence of the contextual integrity undergirding the COVID-19 HP and the filtering by institutional trust, the present study applies a vignette

[8] It should be noted that there is some evidence that trust in government and, to a lesser extent, trust in science have increased in the Netherlands after the first COVID-19 lockdown in March 2020 (Oude Groeniger et al., 2021; Reeskens et al., 2021).

experiment which was fielded 15 months after the outbreak of the COVID-19 pandemic in Europe. It should be noted that an emergency context facilitates the acceptability of novel technologies (Apthorpe et al., 2018). In their vignette study fielded in Germany, Gerdon et al. (2021) found that, pre-pandemic, the transmission of health data to a public institution for a public purpose (i.e., containment of infectious diseases) was the least accepted. However, this changed with the outbreak of COVID-19, which increased the acceptability of the public purpose to share health data (Gerdon et al., 2021). In Poland, however, acceptance of COVID-19 tracking technologies decreased as the pandemic progressed (Wnuk et al., 2021). With this in mind, in the remainder of the section, we describe the informational norms which may be associated with the data exchanges of the COVID-19 HP and formulate hypotheses.

Data recipient

The CI framework states that people's willingness to share information is contingent upon the receiving end (Olson et al., 2005). Previous studies found that surveillance performed by a public entity is more easily accepted than surveillance performed by a private company (Degli Esposti & Santiago Gómez, 2015; van den Broek et al., 2017). Indeed, compared to a private institution, a public institution is more likely to pursue a collective goal, rather than aiming at profit, and can be held accountable in case of data breaches or malfunctions. Consequently, we expect the COVID-19 HP to be more acceptable if the data recipient is a public institution rather than a non-public one [H2].

Information attributes

Previous studies provide descriptive evidence on the acceptability of technologies based on information attributes. In a pioneering study, Olson et al. (2005) showed that participants were unwilling to have their personal email widely distributed, whereas they accepted to have their work credentials disclosed. This indicates that respondents distinguished the acceptability of an information exchange based on the level of disclosure risk, being work credentials – unlike the personal email address - often easily searchable online and displayed on websites.

Applied to the COVID-19 HP, we propose that its acceptability depends on the extent to which it reveals personal information upon identity verification, in two different directions: identification and disclosure risk. On the one hand, the COVID-19 HP may make individuals personally identifiable, e.g. by displaying their full name. On the other hand, the COVID-19 HP may rely on anonymous numeric sequences such as the social security number, which can link to financial statements and other sensitive information on individuals. Consequently, the more invasive the personal information included in the pass, the lower the acceptability of the COVID-19 HP should be [H3], though it remains to explore which of the two characteristics is more problematic between identification and disclosure risk.

Transmission mode

It is also important to consider modes of transmission of data enabled by the COVID-19 HP. One important aspect in this regard is the data retention period. Lewandowsky et al. (2021) found that guaranteeing the deletion of data after 6 months increased the acceptability of COVID-19 app-based tracking technologies. Arguably, this relates to the perception of the risks that the data are re-used beyond the primary function of the COVID-19 HP. The exchange of information via a paper support may be perceived as more clearly confined within the specific purpose of the HP itself, and signal a shorter retention period: for instance, a piece of paper containing the COVID-19 HP QR-code can be shredded after scanning. On the contrary, when the COVID-19 HP is displayed on a digital device, there is a risk of future transmission of data for different, unforeseen purposes, as the digital support also signals a long-lasting retention period, e.g. the CoronaCheck QR-code remains accessible even after scanning. Accordingly, we expect the COVID-19 HP to be more acceptable when it is in the form of a paper certificate rather than on a digital support [H4].

Purpose of use

The ambition of the introduction of a COVID-19 HP is to reopen social life while simultaneously preserving participants' health by excluding individuals without a valid certification. Previous findings suggest that emphasizing prosocial motives compared to self-interest motives may affect the acceptability of the COVID-19 HP. Indeed, some studies found a positive association between prosocial responsibility and the willingness to sacrifice privacy (Kokkoris & Kamleitner, 2020) and adopting COVID-19 tracking technologies (Wnuk et al., 2021), indicating that individuals are more willing to disclose personal information when confronted with a collective public health purpose. Similarly, a recent study found that a stronger intention to adopt protective behaviors against a coronavirus infection followed a message underscoring benefits for others rather than for the self, and was better predicted by the perception of public threat compared to personal threat (Jordan et al., 2021). Accordingly, we would expect that a measure presented as ensuring collective benefit would be deemed more acceptable than a measure aimed at restraining individual participation in social life [H1].

Additionally, different purposes may elicit different informational norms by changing the frame in which norms are entrenched. For instance, Gerdon et al. (2021) found that the acceptability of the data recipient and attributes changed depending on the purpose of data transmission[9], while Apthorpe

[9] More specifically, compared to a company, a public agency was found more appropriate to

et al. (2018) found that the attributes' acceptability was higher when closer to the primary function of the technology considered. In the context of our vignette experiment, we thus expect that the informational norms elicited by the COVID-19 HP have a weaker impact on its acceptability when the measure is justified as collective benefit rather than restraining individual freedom [H5].

Institutional trust and the salience of informational norms

Institutional trust acts as a heuristic in uncertain circumstances by fueling the perception that transmitted data will not be abused (Bradford et al., 2020; Trüdinger & Steckermeier, 2017). With regards to the government, political trust concerns the expectation that power held by political actors will not be abused (Trüdinger & Steckermeier, 2017), thus fostering the acceptability of novel surveillance technologies by reducing the uncertainty created by the data exchange. To give a few examples, opinions towards the police are found to predict the acceptability of privacy-intrusive technologies such as Facial Recognition Technologies in London (Bradford et al., 2020) and surveillance cameras in public places in Russia (Gurinskaya, 2020), whereas political trust is found to enhance support of surveillance policies (Ioannou & Tussyadiah, 2021; van den Broek et al., 2017) and of COVID-19 contact tracing and immunity passport (Altmann et al., 2020; Lewandowsky et al., 2021).

Trust in science can reduce uncertainties by reinforcing the feeling that the measure is warranted to effectively combat the pandemic, and the perception of the benevolent intentions of those who promote the measure. Research on how trust in science affects the adoption of a new technology is scarce; however,

handle data for public purposes compared to private purposes, but only for location and energy data; for health data, a company was more accepted than public agency regardless of the purpose.

a study conducted during the COVID-19 pandemic found that trust in science predicts the adoption of protective measures (Dohle et al., 2020).

Accordingly with the outlined theoretical mechanism and previous findings, we expect a positive association between institutional trust and acceptance of the COVID-19 HP. In addition, we argue that institutional trust affects the privacy evaluations associated to the COVID-19 HP informational norms. By reducing the uncertainty related to the data exchange, the risk of violation associated with the most privacy-invasive features (e.g. non-public data recipient, digital transmission mode) becomes smaller. Therefore, we expect that institutional trust reduces the impact of the informational norms on the acceptability of the COVID-19 HP [H6].

4 • DATA AND METHODS

The acceptability of the COVID-19 HP is investigated by means of a vignette experiment, aligning with other studies that adopted a vignette experimental design in relation to the CI framework (cf. Apthorpe et al., 2018; Gerdon et al., 2021; Horne et al., 2015). The experiment was fielded on the LISS panel, a Dutch probability-based online panel managed by Centerdata. The vignette experiment was part of a series of follow-ups on the coronacrisis in the framework of the Dutch European Values Study (EVS) 2017. The survey and the experiment were administered in Dutch (see Figure B.2 in Appendix B).

While designing the experiment in February–March 2021, we closely monitored the information spread by the news media and the announcements by the Dutch government to make the scenarios realistic. We also tried to avoid too technical details to make it more easily understood by all respondents. Moreover, the question was framed as an hypothetical, describing how the COVID HP "could look like" and not presenting is an actual measure. Ahead of the study, the set up of the experiment was carefully reviewed by the representatives of the LISS panel, and approval was granted by the Ethical Review Board of Tilburg School of Social and Behavioral Sciences.

Vignette design

Respondents were presented with a short text describing a certificate necessary to be granted access to public venues amidst the COVID pandemic. In the description of the COVID-19 HP, we varied its features based on three CI parameters (data recipient, information attributes, transmission principles) and on the purpose of use, for a total of $2x3x2x2 = 24$ different vignettes(see section 2 in Appendix B. Tweaking the features of the HP ensures that we do not measure the political/ideological support for the measure, but evaluate the acceptability of the measure in terms of privacy evaluations.

To test the hypothesis related to the purpose, the vignettes display two conditions, one stressing heightened collective freedom and the other stressing restricted individual freedom. As concerns the comparison between a public vs a non-public data recipient, the vignettes specify that data are managed by either the government or one's own health insurer.[10] With regards to the information attributes, the vignettes present three options of personal data displayed on the pass:

- Initials and partial date of birth, i.e. the minimal amount of information to verify the identity of the pass holder;
- Full name and date of birth, which make a person more easily identifiable though carrying low disclosure risk;

[10] In the Dutch health care system, citizens must provide for their own health insurance (although under some conditions the costs can be refunded by the government). This system makes the health insurance companies credible stakeholder in the HP since they operate within the health-care realm while pursuing their own interest.

- BSN number, i.e. the Dutch social security number, which protects from identification, but has high disclosure potential as it could be used as linkage to other sensitive information.

Finally, for the transmission principle, we distinguished between a digital (app-based or text) and printed support. The theoretical mechanisms, experimental condition and hypotheses are summarized in Table 3.1.

Variables

Aligning with previous studies (Horne et al., 2015; Horne & Przepiorka, 2019; Lewandowsky et al., 2021), we questioned respondents about their willingness to use the COVID-19 HP as a measure of its acceptability, which constitutes the dependent variable in our study. Answer options ranged in a 5-points scale from 'certainly not' to 'certainly yes'. Almost one respondent in two (45.5%) answered they would be certainly willing to use the HP, whereas only one in ten respondents (8.7%) is certainly not willing to (see Figure B.1 in Appendix B).

Institutional trust was measured with the standard EVS question wording, with answer categories ranging from 1 (None at all) to 4 (A great deal). Whereas the government is typically mentioned in the institutional trust scale in the EVS, science has been added to the list of institutions during the COVID-19 EVS follow-up studies.

TABLE 3.1 Summary of experimental conditions and hypotheses.

CI parameter	Theoretical mechanism	Vignette experimental conditions	Hypotheses	
			Acceptance of the COVID-19 HP is higher when...	Conditioned by
Purpose of use	Collective benefit *v.* individual restriction of liberty	A. The HP is aimed at making public places safe for everyone B. The HP is aimed at preventing potentially exposed individuals from participating in social events and travels	H1: ... it is justified for collective freedom purpose	-
Data recipient	Public *v.* non-public institution via orientation towards collective goal	A. Government B. Own health insurer	H2: ...the data recipient is a public authority.	The higher acceptability of a public data recipient, non-invasive information attributes and printed support as transmission principle is relatively weaker when H5: ... the HP is justified for collective freedom purpose H6: ... the subject displays high institutional trust
Information attributes	Identification and disclosure risk	A. Initials and date of birth B. Full name and date of birth C. BSN number	H3: ... the least identifiable/disclosing information is displayed.	
Transmission principle	Risks of extension beyond primary function	A. A code on a piece of paper. B. A code on a mobile app or via text	H4: ... the mode of transmission is on a printed support.	

We control for individual-level variables which may affect the acceptability of the HP. Among the sociodemographic characteristics we control for age, sex and educational level; we also include some pandemic-related attitudes. Unsurprisingly, the perception of COVID risk was found to be positively associated with the acceptance of a contact tracing/immunity passport (Lewandowsky et al., 2021), and we suspect it also provides a justification frame for acceptance of the HP. Consequently, we include a variable measuring concern over coronavirus on a range from 1 (Not at all) to 5 (A lot). Additionally, since the topic of vaccinations is contested despite being a cornerstone of the HP strategy, we also include a measure of vaccination hesitancy, indicated by answering 'probably not' or 'certainly not' to a question on the COVID-19 vaccination intention.

After deleting cases with missing values on one or more variables (84 cases, or 5.5% of the total sample), the analytic sample includes 1,454 individuals. The descriptive statistics are presented in Table 3.2, showing that trust in science is higher than trust in government[11] and that the age distribution in the sample is skewed to the right, with an average age of 59 years. The question wording of all the questions used is included in Appendix B (see Table B.2), alongside a correlation matrix (see Figure B.3).

Analytical strategy

Because of the nature of the dependent variable, we use ordinal logistic regression, using the clm function from the ordinal package (Christensen, 2019) in Rstudio version 1.4.1106. We used Partial Proportional Odds models (PPO) (O'Connell, 2011) following the results of a likelihood ratio test (see Ta-

[11] Trust in government and trust in science are, unsurprisingly, positively correlated with each other, but the strength of the correlation is moderate ($\rho = 0.35, p < 0.001$) and allows to treat them separately.

TABLE 3.2 Descriptive statistics.

Statistic	Min	Max	Mean	St. Dev.	N
Willingness to use HP	1	5	3.92	1.27	1,454
Trust in government	0	3	1.36	0.71	1,454
Trust in science	0	3	2.22	0.63	1,454
Concern over coronavirus	0	4	2.37	0.86	1,454
Vaccination hesitant	0	1	0.09	0.29	1,454
Female	0	1	0.51	0.50	1,454
Educational level	1	6	3.87	1.50	1,454
Age	17	95	58.45	16.77	1,454

ble B.1 in Appendix B). Accordingly, the coefficients of trust in government are allowed to vary for each category of the dependent variable. For ease of presentation, the results will be mostly presented in terms of predicted probabilities (and related 95% confidence intervals).

5 • RESULTS

HP Acceptability

First, a model is estimated to evaluate the general willingness to use the HP (see model 1 in Table 3.3). Both trust in government and trust in science have a positive association with the acceptability of the COVID-19 HP, as illustrated also in Figure 3.1. For instance, the probability of accepting the HP (category 'Certainly yes') among those who trust the government is 16 percentage points higher compared to those who do not trust it, whereas when differentiating on the level of trust in science, the acceptability grows by 24 percentage points. For those displaying high trust in either institution, the

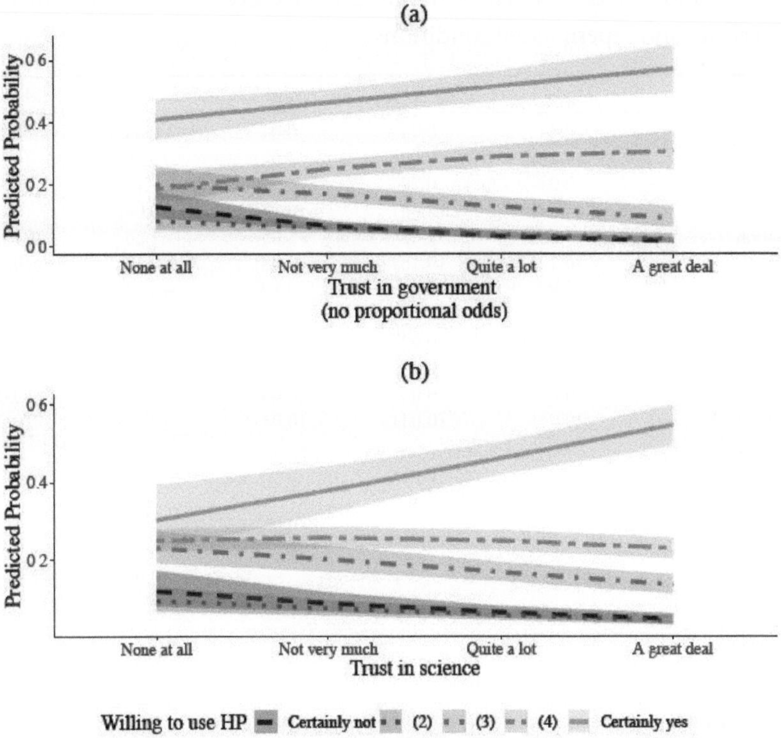

FIGURE 3.1 Predicted Probabilities (with 95% Confidence Intervals) of Willingness to use COVID-19 HP by institutional trust (PPO model), N = 1,464.

probability of refusing to use the HP (category 'Certainly not') is almost null and more than 10 percentage points lower compared to those who do not trust.

With regards to control variables, we found that the HP is more acceptable among those who are concerned about the coronavirus, those who are not hesitant about getting vaccinated against COVID-19, elderly people and men. Educational attainment is not significantly related to the acceptance of the COVID-19 HP.

TABLE 3.3 PPO models of Willingness to use the COVID-19 HP by individual characteristics and experimental conditions.

	Model 1	Model 2	Model 3	Model 4	Model 5
Trust in science	0.34***	0.34***	0.34***	0.34***	0.34***
	(0.09)	(0.09)	(0.09)	(0.09)	(0.09)
Concern over coronavirus	0.22***	0.22***	0.22***	0.22***	0.22***
	(0.06)	(0.06)	(0.06)	(0.06)	(0.06)
Vaccination hesitant	−1.66***	−1.66***	−1.65***	−1.66***	−1.66***
	(0.18)	(0.18)	(0.18)	(0.18)	(0.18)
Female	−0.22*	−0.22*	−0.22*	−0.22*	−0.23*
	(0.10)	(0.10)	(0.10)	(0.10)	(0.10)
Age[a]	0.02***	0.02***	0.02***	0.02***	0.02***
	(0.00)	(0.00)	(0.00)	(0.00)	(0.00)
Educational attainment[a]	0.04	0.04	0.04	0.04	0.04
	(0.04)	(0.04)	(0.04)	(0.04)	(0.04)
Experimental conditions					
Purpose: ensure safe access		0.01			
		(0.10)			
Recipient: own insurer			−0.05		
			(0.10)		
Support: digital				−0.06	
				(0.10)	
Attribute: full name					−0.01
Attribute: BSN					−0.17
					(0.12)
Thresholds: Trust in government					
1|2	−0.80**	−0.79**	−0.82**	−0.82**	−0.86**
	(0.28)	(0.29)	(0.29)	(0.29)	(0.29)
2|3	−0.21	−0.20	−0.23	−0.23	−0.27
	(0.27)	(0.28)	(0.27)	(0.27)	(0.28)
3|4	0.75**	0.76**	0.72**	0.72**	0.68*
	(0.26)	(0.27)	(0.26)	(0.26)	(0.27)
4|5	1.50***	1.50***	1.47***	1.47***	1.43***
	(0.26)	(0.27)	(0.26)	(0.26)	(0.27)
Thresholds: Trust in government					
1|2	−0.75***	−0.75***	−0.75***	−0.75***	−0.75***
	(0.15)	(0.15)	(0.15)	(0.15)	(0.15)
2|3	−0.64***	−0.64***	−0.64***	−0.64***	−0.64***
	(0.12)	(0.12)	(0.12)	(0.12)	(0.12)
3|4	−0.52***	−0.52***	−0.52***	−0.52***	−0.52***
	(0.09)	(0.09)	(0.09)	(0.09)	(0.09)
4|5	−0.22**	−0.22**	−0.22**	−0.22**	−0.22*
	(0.08)	(0.08)	(0.08)	(0.08)	(0.08)
Model fit					
AIC	3736.40	3738.39	3738.11	3737.99	3737.99
BIC	3810.35	3817.62	3817.34	3817.22	3822.50
Log Likelihood	−1854.20	−1854.20	−1854.06	−1854.00	−1852.99
Num. obs.	1454	1454	1454	1454	1454

Log odds (standard error in parenthesis); [a] Median-centered

$^{***}p<0.001$; $^{**}p<0.01$; $^{*}p<0.05$;

Informational norms

The results presented in model 2 to model 5 in Table 3.3 include the experimental conditions, and indicate that the willingness to use the HP does not depend on the informational norms associates with the HP (see Figure 3.2): when varying the purpose of the pass, the data recipient, and the transmission principle, the predicted probabilities of selecting each category remain substantially the same across experimental settings. Only the BSN, the data attribute with the highest disclosure risk, seems to encounter more resistance, but the difference in acceptability with the other attributes is not statistically significant. Hypotheses 1 to 4 are hence to be rejected.

In order to test hypothesis 5, we added an interaction between the purpose of use of the HP with each of the other three CI parameters, expecting a reduced salience of the latter when the stated purpose was ensuring safe access to social venues. This is not the case, as the impact of each parameter on the willingness to use the HP does not vary according to the purpose of access (see Figure 3.3 and Table B.3). If anything, differences between experimental conditions seem to be larger rather than smaller when the purpose is presented as 'ensuring safe access': for instance, while the probability of being willing to use a pass 'preventing risky access' is 2.4 percentage points lower when the data recipient is the government, when the pass is presented as 'ensuring safe access' the same probability is 5 percentage points higher when the government is involved. However, these differences are not statistically significant, leading to reject hypothesis 5.

Informational norms by individual predispositions

Finally, hypothesis 6 is tested by interacting the CI parameters - data recipient, support and attributes – with individuals' trust in government and trust in science. The results are reported in terms of predicted probabilities of select-

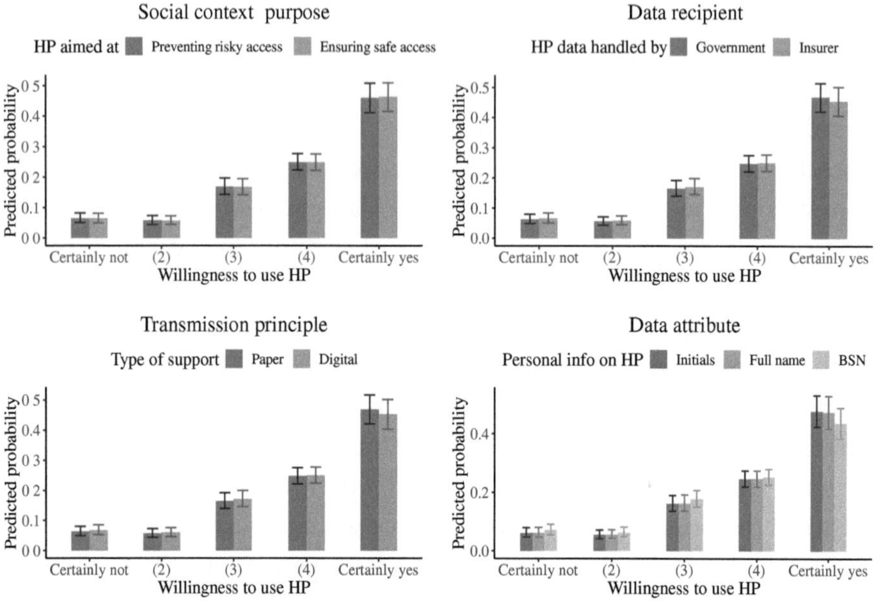

FIGURE 3.2 Predicted Probabilities of Willingness to use HP with 95% Confidence intervals by experimental conditions, estimated via ordinal logistic model (PPO model), N = 1,454.

ing the two extreme categories in Figure 3.4 and Figure 3.5 for the ease of presentation (see Table B.3 for full results).

Contrary to the hypothesis, trust in government does not have an impact on the informational norms associated with the HP, as neither the data recipient nor the transmission principle display different associations with the willingness to use the HP depending on the level of trust in government. Strikingly, the probabilities of accepting an insurer as data recipient is larger at high levels of trust in government, whereas the acceptability of the government as data recipient remains stable; yet, the difference in the slopes is not statistically sig-

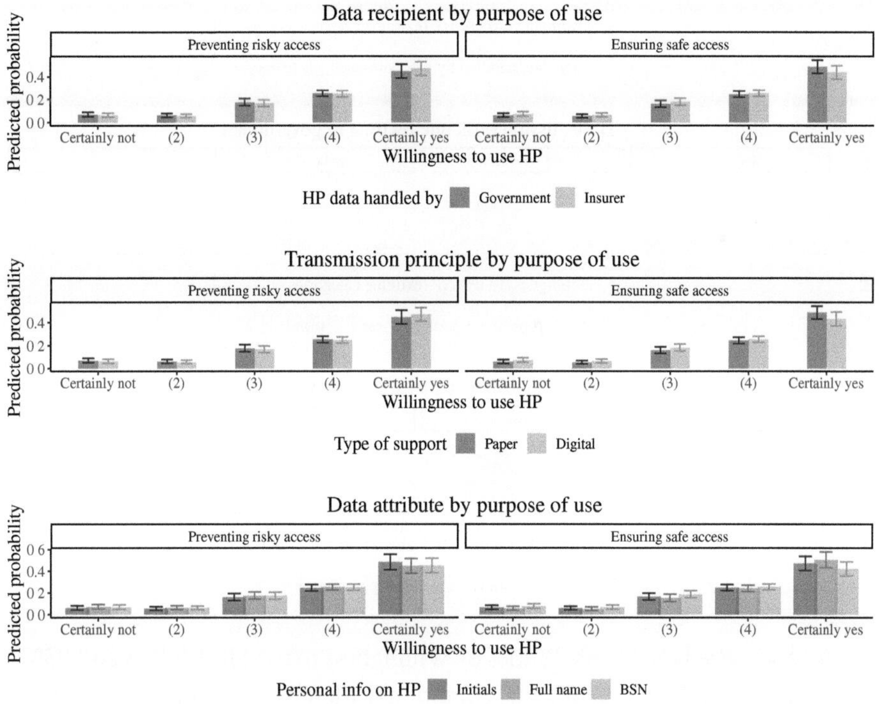

FIGURE 3.3 Predicted Probabilities of Willingness to use Health Pass and 95% Confidence intervals by experimental conditions by the purpose of use of the HP, estimated via ordinal logistic model (PPO model), N = 1,454. Full models available in Appendix B

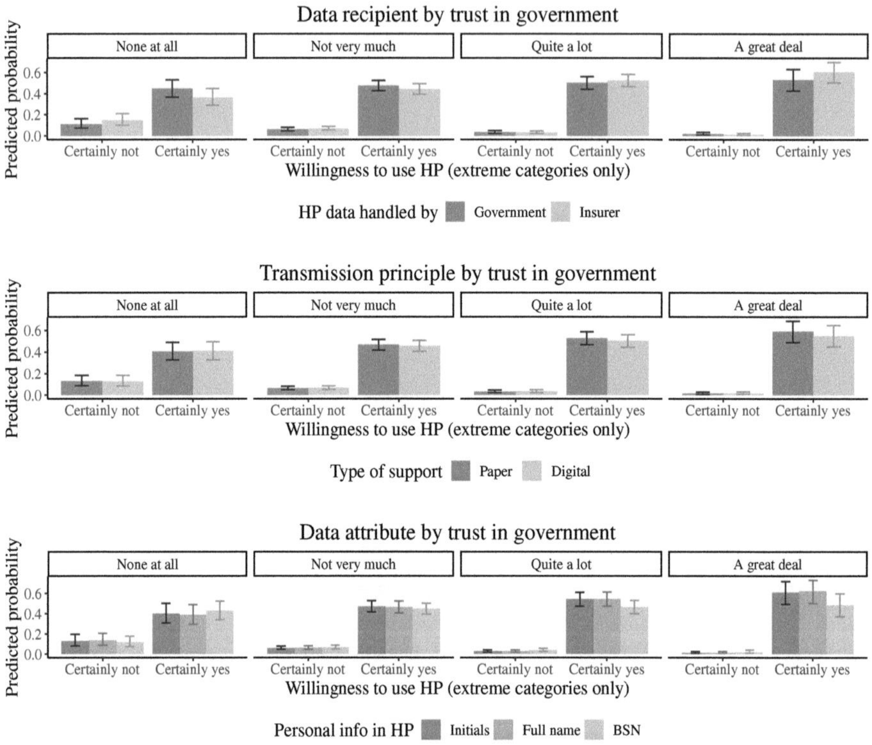

FIGURE 3.4 Predicted Probabilities of Willingness to use Health Pass and 95% Confidence intervals by experimental conditions by levels of trust in government, estimated via ordinal logistic model (PPO model), N = 1454. Only extreme categories are displayed for parsimony, full results can be found in Appendix B

nificant. As for data attributes, the acceptability of the BSN appears to grow more slowly compared to the other attributes at higher levels of trust, but the difference is not statistically significant.

Turning to trust in science, the relative increment in acceptability of the own insurer as data recipient compared to the government increases significantly across levels of trust in science ($p < 0.05$) (see Figure 3.5): whilst at the lowest level of trust in science the insurer is 17 percentage point less acceptable than the government, at the highest level of trust in science the two are almost equally accepted (both with a probability higher than 0.50). The BSN is relatively less accepted than other data attributes at high levels of trust, but the difference is not statistically significant. The impact of the type of support on the willingness to use the HP remains stable at different levels of trust in science.

6 • CONCLUSION AND DISCUSSION

In this study, we adopted the Contextual integrity framework (Nissenbaum, 2010) and fielded a vignette experiment to investigate whether informational norms engrained in specific features of a COVID-19 HP, in combination with individual predispositions, affect the acceptability of such a health surveillance system among a sample of Dutch residents.

We found that in May 2021 there was overall large support for a COVID-19 HP among Dutch citizens, and that this support is not eroded by more intrusive privacy features. Through the lenses of the CI framework, according to most Dutch people, the HP complies with the expectations pertaining the exchange of information in the context of a public health emergency. The majority of the respondents is willing to use such a tool notwithstanding whether data is managed by a legitimate public authority such as the government or by an insurance company (unlike suggested by previous studies, cf. Degli Esposti & Santiago Gómez, 2015; van den Broek et al., 2017), whether the pass is transmitted on paper or digitally, and whether the information used to verify the pass is potentially revealing of one's identity.

The salience of these informational norms for the HP's acceptability was not even noticeable when the pass was presented a freedom-restraining tool, aimed

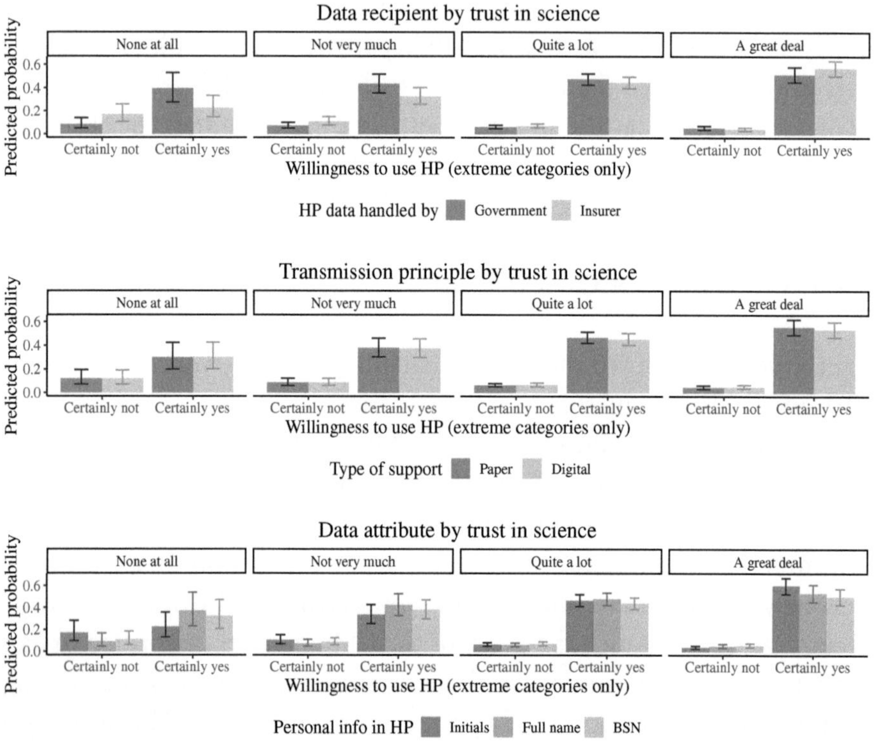

FIGURE 3.5 Predicted Probabilities of Willingness to use Health Pass and 95% Confidence intervals by experimental conditions by levels of trust in science, estimated via ordinal logistic model (PPO model), N = 1454. Only extreme categories are displayed for parsimony, full results can be found in Appendix B

at preventing potentially exposed individuals from participating in events. In hindsight, this type of framing may have underlined the risks of infection making it more evident, and thus strengthening the acceptability of a potential privacy invasion.

More generally, one potential explanation for the lack of impact of the privacy-intrusive features of the HP on its acceptability may be related to the emergency situation, in line with alternative privacy approaches such as the privacy-security trade-off, according to which increased security trumps privacy concerns. Yet, the positive impact of the concern for the coronavirus and the willingness to get the COVID-vaccination on the acceptance of the HP (cf. also Lewandowsky et al., 2021) suggests a context-specific evaluation of costs and benefits associated with the use of the pass. Concerned respondents, eager to safely resume social life after months of restrictions, are more willing to enable the exchange of personal information on themselves and their health status. This aligns with the privacy calculus perspective, according to which people undertake privacy decisions based on a rational evaluation of costs and benefits (Hallam & Zanella, 2017). This perspective also confirms the importance of the context for privacy decisions, as the acceptability of the HP in our study occurs within a specific period – the COVID-19 pandemic – and may yield different results in a different moment in time, as showed by the study of Gerdon et al. (2021).

Informational norms appear not to be filtered by individual predispositions towards the institutions deploying the HP, as there is only limited evidence that institutional trust buffers the differences in acceptability as a function of the features of the pass. At high levels of trust in science, there is less differentiation in acceptability of the government *v.* one's own insurer as data recipients, driven by a steeper growth in the acceptability of the latter across levels of trust. Hence trust in science seems to spill over to other institutions and increase their acceptability as data recipients regardless of their mission. Yet, the acceptability of the HP based on the transmission principle and data attributes does not vary across levels of institutional trust.

Nevertheless, the acceptability of the COVID-19 HP itself is affected by institutional trust, as those who trust the government and/or science are clearly more prone to accept the HP, aligning with previous studies on the positive impact of institutional trust on the acceptability of privacy-intrusive measures (Altmann et al., 2020; Bradford et al., 2020; Lewandowsky et al., 2021; Trüdinger & Steckermeier, 2017). Trust in the two institutions hereby considered seem to work in similar ways, in line with previous studies on institutional trust suggesting that respondents do not really differentiate among institutions (cf. Hooghe, 2011). Interestingly, however, they maintain a distinct direct impact on the acceptability of the COVID-19 HP independently of each other.

Our study presents some limitations. First, the timing of the survey (May 2021) overlapped with news on the EU Digital Green Pass and the national implementations. While this may have enhanced the respondents' understanding of the vignette, it may also have rendered some of the scenarios unrealistic (e.g., data handled by own insurance company). Second, the study was conducted in the Netherlands, a country with high levels of digitalization also in the healthcare realm. Accordingly, Dutch people, already accustomed to, e.g., retrieving test results from online systems, may be overall less concerned about sharing health-related data, and less sensitive to the features of such a system. Finally, we tried to incorporate different informational norms in the varying features of the HP, but some elements may have been too implicit. Future studies should find ways to better capture the normative aspects of information exchanges.

It is important to point out that the response to the COVID pandemic has been politicized, hence also the support for the HP may be rooted in ideological stances rather than on the rational evaluation of the potential privacy violations associated with it. While this aspect should be explored further in the context of the COVID-19 pandemic, some preliminary insights are offered in

Chapter 5, where the role of authoritarian attitudes is investigated in relation to the educational gradient in acceptance of surveillance.

There are two main implications to our study. First, those who do not support current political institutions risk to remain even more marginalized in the post-pandemic social life. Considering that those who refuse the HP are also those who may refuse to get vaccinated against COVID, this poses a social as well as a public health problem, and requires extensive research. Second, many scholars and activists warn again surveillance creep: the finding that the emergency situation justifies the adoption of the pass creates an important precedent, and warrants attention.

The Closing Educational Gap in E-privacy Management in European Perspective

ABSTRACT

Educational gaps are increasingly salient as skills and knowledge gain prominence in digital societies. E-privacy management, namely the ability to control the flow of information about the self, is an important asset nowadays, since a skillful use of digital technologies enables full participation in social life and limits the exposure to unwarranted algorithmic processes. We investigate whether and why education affects e-privacy management, and whether the educational gaps vary following a country's degree of digitalization. We empirically test two sets of mechanisms, one derived from the digital divide and diffusion of innovations theories, the other from the reflexive modernization theory. The study employs Eurobarometer 87.1 data (N = 21,177), collected in 2017 among representative samples from 28 European countries, and uses multilevel linear regression model. Findings suggest that the years spent in education positively affect e-privacy management, and that this effect is largely mediated

This chapter, with slight differences, has appeared in print. See Maineri, A., Achterberg, P., & Luijkx, R. (2021). The closing educational gap in e-privacy management in European perspective. *Sociological Research Online.* https://doi.org/10.1177/13607804211023524. Supplementary materials available on the journal's website, replication materials available on OSF (https://doi.org/10.17605/OSF.IO/CTFX8).

by digital skills and internet use, and to a lesser extent by a reflexive mindset. The educational gap in e-privacy management narrows in more digitalized countries.

1 • INTRODUCTION

Due to the centrality of knowledge and information in contemporary societies, education has become a powerful indicator of social position (Bovens & Wille, 2017). Economic theories, e.g. the skill-biased technological change, explained how technological change aggravates inequalities by taking over tasks from the unskilled workers and favoring workers with higher skills (Acemoglu, 2002). At the individual level, the complexity of the technologies that increasingly mediate daily lives is more easily handled by more educated people (Cruz-Jesus et al., 2016). The educational inequalities arising in the digital, information-intense environment hence become important factors in the reproduction of social inequalities in contemporary societies, as also elaborated in Chapter 1. In this study, we analyze a potential expression of educational inequalities in digital societies, namely the educational gap in e-privacy management and its configuration in European countries.

Privacy describes the boundaries between the self and society (Anthony et al., 2017; Marx, 2016), and plays an important role for social order by involving monitoring and social control (Anthony et al., 2017). Broadly intended, privacy means 'the access of one actor (individual, group, or organization) to another' (Anthony et al., 2017, p. 251). Privacy decisions depend upon the context (Nissenbaum, 2010; Y. J. Park & Shin, 2020), e.g. who is going to have access and for what purposes, and privacy norms define what is appropriate in terms of access in different situations (Anthony et al., 2017). One's ability to control access to the self as well as the capability to access others is defined as privacy management (Anthony et al., 2017). For individual citizens, (information) e-privacy management concerns the control of the flow of information about the self that is released online (Blank et al., 2014; Cho & Larose, 1999; Kokolakis, 2017; Y. J. Park, 2015): it is not about releasing information *per se*, but about knowing what information is released, to whom, and for which purposes. E-privacy is not uniquely rooted in the digital sphere, since the large amount of personal information exchanged online as well as the far-reaching consequences of a breach of privacy online make it a key aspect of general privacy protection nowadays.

While previous studies often focused on disclosure behaviors and/or management of privacy settings on Social Networking Sites (Bartsch & Dienlin, 2016; Boyd & Hargittai, 2010; Debatin et al., 2009; Litt, 2013a; Litt & Hargittai, 2014; Y. J. Park, 2018), in this study, inspired by the approach of Büchi et al. (2017), we focus on e-privacy management within general internet use.

One's e-privacy management is increasingly challenged in the digital society as individual characteristics and preferences are monitored and reproduced via computational processes (see Chapter 1 for a more extensive discussion on this). First of all, in their daily online interactions internet users need to release information about the self in exchange for services (Acquisti et al., 2015; Kokolakis, 2017; van Dijck, 2014). This exchange makes data released online a valuable asset, which users rarely acknowledge. For example, social media platforms such as Facebook, do not directly charge a fee to users, yet generate revenues by selling targeted advertisement space based on the elaboration of users' data. Secondly, digital technologies enable information collection at little cost and without the monitored subjects necessarily being aware of it. Finally, as more and more information on the self is shared online and easily harvested, there are growing risks of abuse. On the one hand, cybercrime is widespread, but, since it often does not have immediate repercussions on victims (e.g. identity theft), it goes unnoticed and underreported. On the other hand, concerns over discrimination and social sorting (Acquisti et al., 2015; Anthony et al., 2017; Lyon, 2005; Mann & Matzner, 2019) are growing as decision-making processes are increasingly delegated to algorithms.

E-privacy management is a critical resource in contemporary digital societies (Büchi et al., 2017), and as such it is likely to be unevenly distributed, with the risk of leaving some social groups vulnerable to the negative consequences of digitalization (Lupton, 2016). While this is generally attributed to resource constraints (Anthony et al., 2017) or socialization processes (Y. J. Park, 2018), the mechanisms explaining such unequal distribution are not clear. We draw on literature on the digital divide and diffusion of innovation, and on reflexive modernization, in order to explain the educational gap in e-privacy management. Whereas the former theory refers to internet use and digital skills as unevenly distributed resources that could explain differences in e-privacy management, the latter pinpoints the role of knowledge and risks awareness in an increasingly information-intense environment.

Critical resources for a fruitful use of digital technologies tend to be associated with education (Cruz-Jesus et al., 2016). Previous findings on the relationship between education and e-privacy management are mixed. Whereas some studies did not find any significant educational differences in e-privacy control (Cho et al., 2009; Litt, 2013a; Y. J. Park, 2011, 2013), Y. J. Park and Chung (2017) highlighted how education positively affects awareness of, interest in, and control of privacy online. We therefore attempt at clarifying this discrepancy, and ask whether and why the higher educated are better equipped in managing their privacy online than the lower educated. By answering this question, we shed light on new potential stratification mechanisms taking place in the digital society.

Secondly, we expand our focus compared to Chapters 2 and 3 and look at the comparison across European countries which are at different stages of the digitalization process, in order to evaluate whether the digital (infra)structure in a country leads to a widening or narrowing of the educational gap in e-privacy. We ask whether the degree of digitalization affects the educational gaps in e-privacy management. Comparing countries with different degrees of digitalization may help forecasting trends over time, as digitalization processes expand globally.

By engaging with the study of educational gaps in the management of privacy online in comparative perspective, we aim at enriching the insights on e-privacy and on educational inequalities in the digital society. We examine and empirically test two mechanisms that could underlie social inequalities in the management of e-privacy. Although the two perspectives share expectations concerning the educational gap in e-privacy management, they suggest different mechanisms, as well as different trajectories of educational gaps according to the degree of digitalization of the country. Analyzing inequalities in e-privacy management also aims at informing policy makers, as the European Commission's Digital Single Market initiative aims at ensuring that European citizens can fully profit from the opportunities offered by digitalization.

2 • THE EDUCATIONAL GAP IN E-PRIVACY MANAGEMENT

Diffusion of innovation and digital divide

At the individual level, the existence of digital divides, i.e. the inequalities in the access to, use of and gains from the internet has been largely documented (Hargittai, 2002; Lutz, 2019; Robinson et al., 2015; Scheerder et al., 2017; van Deursen & Helsper, 2015; van Dijk, 2005, 2013). Offline resources determine the extent to which one has access – broadly intended – to digital technologies. It is a cumulative model: when the access gap ('first-level digital divide') is overcome, a skill- and use- divide emerges ('second-level digital divide'); as the skill divide closes, a gap in the benefits obtained by internet use arises ('third-level digital divide'). Sources of online divides align with traditional, 'offline' sources of inequalities: educational level, occupational prestige, gender, age, etc. (for extensive reviews of different levels of the digital divide, see Lutz, 2019; Scheerder et al., 2017).

E-privacy management is a critical skill in the digital era (Büchi et al., 2017; Y. J. Park & Chung, 2017). Privacy is related to power relationship and hence unequally distributed within societies; privacy management additionally requires skills and resources to be enacted (Anthony et al., 2017; Büchi et al., 2017). A Swiss study found that privacy protection is mostly explained by internet skills, and, to a lesser extent, by privacy concerns, and that a more intense internet use indirectly affects privacy protection by increasing exposure to privacy breach (Büchi et al., 2017). Y. J. Park (2011, 2013) found a positive effect of knowledge and familiarity with the internet on privacy control among US citizens, while Bartsch and Dienlin (2016) found a positive association between time spent online and e-privacy literacy in Germany. In other words, e-privacy management tends to be prevalent among groups that are on average more skilled and familiar with digital technologies, along the lines of the digital divide argument. Hence, we expect that the higher educated manage their privacy online more than the lower educated due to their higher frequency of internet use [Hypothesis 1a] and their stronger digital skills [Hypothesis 1b].

Digital divides follow the model sketched by Rogers (2003) in his theory on the diffusion of innovations. Accordingly, an innovation (e.g., a new technology) is progressively adopted by five groups of people, ranging from the first to adopt the innovation to the last ones: innovators, early adopters, early majority, late majority, and laggards. These groups display systematic differences (Neuman et al., 2011), and early adopters tend to be more educated and be in higher social strata compared to late adopters (Rogers, 2003), therefore generating inequalities such as the digital divide. However, as the innovation is adopted over time by an increasingly larger portion of the population, socio-economic factors become weaker predictors of an innovation's adoption. This theory would explain why digital divides - as the educational gap in e-privacy management - should be smaller in countries that have a higher degree of digitalization: in countries where a large portion of the population is interested by technological developments, the lower educated are likely to catch up with a skillful use of technologies. We hence expect that the positive effect of education on e-privacy management is weaker in more digitally advanced societies [Hypothesis 2].

Digital risks and Reflexive Modernization

According to Beck (1992), as technology advances, the more hazards and insecurities come with it. Hence, inextricably linked to modernization is risk – a 'systematic way of dealing with hazards and insecurities induced and introduced by modernization itself' (Beck, 1992, p. 21). Along these lines, the emergence of digital technologies, while offering many potential benefits and possibilities, produces new harms that can quickly escalate due to the pervasiveness in people's daily lives (Lupton, 2016). Unlike traditional risks (e.g. natural catastrophes), people tend not to be physically damaged by digital risks (Beck, 2013), which makes their timely acknowledgment even more critical.

In modern societies, risk is a relevant element in stratification processes (Beck, 1992), for many reasons. First, risks affect some groups in society more than others, and this may run along 'traditional' social-class lines. Moreover, risks depend upon knowledge about them, and knowledge is not equally distributed within societies, hence it follows that those who are aware of risk are those who are more educated and/or informed (Beck, 1992).

In order to understand how people cope with and analyze risk we use the theory of reflexive modernization (Beck, 1992). This theory describes how modern societies started to question their own modern advances and technological solutions. The central claim is that in modern countries there is a strong emphasis on the idea that the typical risks in these societies are 'manufactured' or produced by modern technological innovations. Pointing to newly emerging, and largely unforeseen risks, in reflexively modern societies, people perceive these modern solutions as sources of newly emerging problems. Reflexively modern individuals soon start to analyze all risks they face, including those potentially produced by technological innovations. In these reflexively modern societies, it is implausible that technology is embraced without reservation (Lupton, 1997). While societies increasingly rely on scientific and technological advances, science and technology become also increasingly targets of reflexive criticism (Nettleton & Burrows, 2003; Price & Peterson, 2016).

Theorizing on reflexive modernization emphasizes two aspects. First, the advancement of information and knowledge is one of its central features (cf. Makarovs & Achterberg, 2017; Price & Peterson, 2016). Because of the widely accessible information and knowledge, people can analyze the unintended and latent consequences of technological interventions. Second, some individuals in these societies are better able to analyze the risks brought by technological innovations (Achterberg et al., 2017; Makarovs & Achterberg, 2017). Accordingly, it can be expected that, because of their reflexive mindset, and their cognitive abilities, the higher educated are more concerned about the introduction of newly emerging technologies and are more inclined to actively protect their e-privacy. We hence expect that the higher educated manage their privacy online more than the lower educated because of their reflexive mindset [Hypothesis 3].

As the technological advancements progress alongside hazards, risk becomes increasingly salient as a stratification mechanism, and modern Information and Communication Technologies (ICTs) constitute the very infrastructure for the reflexive attitude to flourish (Nettleton & Burrows, 2003). Consequently, the reflexive attitude becomes a stronger driver of inequalities in online privacy management in countries that are at a later stage of digitalization. We hence expect that the positive effect of education on e-privacy management is stronger in more digitally advanced societies [Hypothesis 4].

3 • DATA AND METHODS

In order to address the research questions and test the hypotheses, we used data from the Eurobarometer 87.1 (European Commission and European Parliament, 2017) which investigated e-privacy- and cybersecurity issues. The questionnaire was fielded in 28 European Union Member states in 2017 via face-to-face Computer Assisted Personal Interviews (CAPI) to representative samples of the population above 15 years old.

Respondents not using the internet were not asked questions about e-privacy, hence our study is restricted to internet users only. In Appendix C we address the individual characteristics of those not using the internet and provide a brief description of how prevalent they are in each country.

Individual-level variables

In order to measure *E-privacy management*, respondents indicated whether in the last three years, because of security and privacy issues[1], they performed a series of actions, listed in Figure 4.1. Only one out of ten respondents had ever cancelled an online purchase, and more than two out of five respondents mentioned they installed or changed anti-virus software.

In order to create a single measure of e-privacy management, we performed a principal components factor analysis on the matrix of tetrachoric correlations among the dichotomous items. We adopted an exploratory approach, since – to our knowledge - the quality of this measurement instrument has not been previously assessed. We excluded the three items ('Less online purchases', 'Less online banking', and 'Only use own computer') that performed poorly[2] with the others. This choice is also justified by a critical assessment of the content, as these three items refer to general actions, in contrast with the remaining items

[1] Exact wording of question QD17 'Among the following possible actions you might have undertaken in the last three years because of security and privacy issues when using the Internet please select those that apply to you?' – multiple answers.

[2] The three items displayed factor loadings below 0.40 and, the uniqueness (unexplained variance) was 0.84 and above

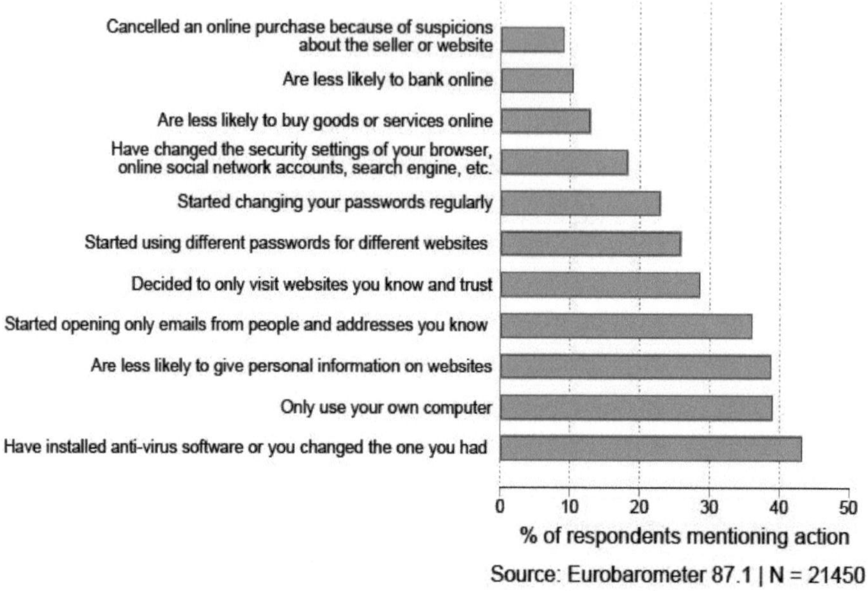

FIGURE 4.1 Percentage of mentions for each e-privacy management action.

that explicitly mention privacy-related actions, e.g. changing passwords or disclosing personal information. Eight items with factor loadings > 0.5 on the first factor (which explains 43% of the variance) were retained (see Table 4.1). The eight items formed a sufficiently reliable index (Kuder–Richardson Formula 20 = 0.66). Although the items mostly refer to a specific computer-related use of the internet which, with the spread of mobile devices, is not completely up-to-date, this measure is comparable to one of the dimensions of e-privacy behaviors Cho et al. (2009) found, and labelled as proactive protecting behaviors. An unweighted sum score was computed on the eight items, and correlated strongly ($rho = 0.99$) with the factor score. See Table 4.2 for the descriptive statistics of this variable.

TABLE 4.1 Factor loadings and uniqueness of principal components factor analysis (N = 21,450).

Variable	Factor	Uniqueness
Less personal information	0.67	0.55
Security settings	0.70	0.51
Only trusted websites	0.52	0.73
Different passwords	0.75	0.44
Reject unknown emails	0.69	0.51
Anti-virus software	0.71	0.49
Cancel online purchase	0.55	0.69
Regularly change passwords	0.60	0.63
Explained variance	43%	

Education is based on the age at which full time education was completed.[3] For those who are still studying, the age at finishing education was imputed as the current age . We subtracted the age at which school starts in each country (cf. Sharp, 2002), which resulted in a measure of years spent in full-time education. Respondents marked up to 79 years spent in full-time education, which is an extremely high value probably due to a misinterpretation of the question. The value corresponding to the 95th percentile, 22 years, is a good approximation for someone who has reached a PhD degree. Hence, we truncated the distribution and assigned the value of 22 years spent in education to all values above the 95th percentile. In this way the range of years spent in education was more plausible, and fewer respondents were excluded from the analyses. Figure 4.2 displays the distribution of the years spent in education across countries and shows that, on average,

[3] A measure of the highest attained level of education was not available

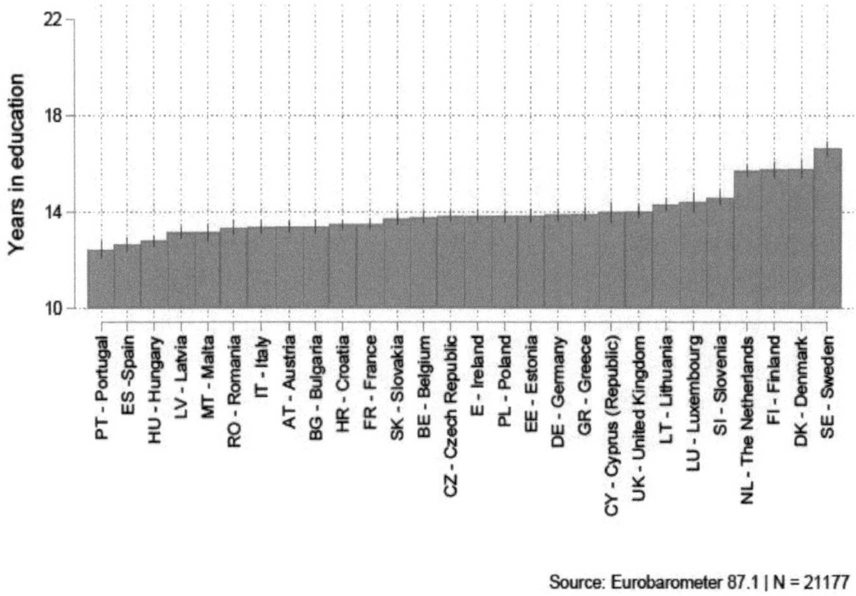

FIGURE 4.2 Mean of years spent in education by country (error bars represent 95% confidence intervals).

respondents from Northern European countries tend to spend more time in education than respondents in Southern countries, aligning with the statistics on tertiary education attainment in Europe (Eurostat, 2020).

As for *internet use*, an index measuring the frequency of performing several activities of the internet (e.g. use internet at home, at work, using social media, etc.) was available in the dataset. After excluding those never using the internet, the index consists of five categories, ranging from '(Almost) everyday' to 'Less often (than two/three times a month)'. Values were recoded so that high values indicate more internet use. We measured self-reported digital skills by the question 'You consider yourself to be sufficiently skilled in the use of digital technologies...' for several domains: in one's daily life, to do a job, to do a future job, to use online public services and to benefit from digital learning opportunities. Answers ranged on a four-point scale from 'totally agree' to 'totally disagree'. The items form a reliable scale (Cronbach's alpha = 0.90), and a principal components factor analysis revealed one factor capturing 73% of the variance. Some of the items were only administered to selected respondents (e.g. 'skills in one's job' are only measured among those who are currently employed), hence average scores on the items were computed taking into account only the items administered to each respondent.

We used a dichotomous item measuring the belief that it is the respondent's personal responsibility to tackle climate change as an indicator of a *reflexive mindset*. The question taps on the acknowledgment of the man-made nature of contemporary risks. Even though the focus on climate change only is suboptimal, data limitations did not allow for a more refined measurement. An alternative measure for this concept is shown in Appendix C and yielded similar results.

In the analyses we controlled for socio-demographic characteristics that are usually associated with e-privacy management, such as gender, age, unemployment status, being a student and the settlement type. Age is particularly relevant because previous studies found that e-privacy management is more common among young people (Blank et al., 2014; Litt, 2013a). Findings on gender are mixed: whereas Park 2011, 2015 found that men are more protective, Litt (2013a) found that women tend to enact more diverse privacy management behaviors on SNSs. Finally, e-privacy management is also common practice among university students, despite a general feeling of inevitability of data breach (Hargittai & Marwick, 2016). See Table 4.2 for the descriptive statistics of all variables used in the models.

In the analyses, all continuous independent variables have been centered around the grand-mean (cf. Hox, 2002, pp. 54–58). We deleted the missing val-

TABLE 4.2 Descriptive statistics of individual characteristics (N = 21,177).

Variable	Mean	Std. Dev.	Min	Max
E-privacy management	2.25	1.91	0	8
Years spent in full–time education	14.09	3.92	0	22
Digital skills	3.13	0.77	1	4
Internet use	3.77	0.70	0	4
Variable	**Proportion**		**Min**	**Max**
Climate change own responsibility	0.27		0	1
Unemployed	0.07		0	1
Student	0.08		0	1
Female	0.54		0	1
Age				
15–24	0.11		0	1
25–34	0.17		0	1
35–44	0.19		0	1
45–54	0.19		0	1
55–64	0.17		0	1
65–74	0.13		0	1
75+	0.04		0	1
Settlement type				
Rural area or village	0.31		0	1
Small or middle sized town	0.40		0	1
Large town	0.29		0	1

ues listwise, leaving 21,177 observations, nested in 28 EU countries. Countries' sample sizes range from 310 respondents in Malta to 1,187 in Germany.

Country-level variables

For the degree of *digitalization* we used the Digital Economy and Society Index (DESI). The DESI is developed within the context of the Digital Single Market initiative of the European Commission. This index considers the countries' digital performance in five different areas: connectivity, digital skills, internet use, integration of digital technologies and digital public services.[4] Consequently, the DESI does not only reflect the prevalence of internet use among the population, but it captures the permeation of the digital services among private enterprises and public institutions. Higher scores stand for more digitalization of a country. We retrieved results from 2017 to match the data collection period of the Eurobarometer. The DESI ranges from 31.9 (Romania) to 65.6 (Denmark) and is shown in Figure 4.3. Per capita GDP in Purchasing Power Standards for 2017, retrieved from Eurostat (reference: PRC PPP IND), is used as a country-level control variable. This metric expresses a country's per capita GDP in relation to the EU average GDP per capita (= 100). The variable ranges from 49 (Bulgaria) to 253 (Luxembourg), with a mean of 99.9 and a standard deviation of 41.2. For the purposes of the analyses, the DESI and GPD have been centered around the mean.

Analytical strategy

In order to test the hypotheses, we estimated multilevel linear regression models, with 21,177 respondents nested in the 28 EU countries. First, we fitted an

[4] Retrieved from https://ec.europa.eu/digital-single-market/en/news/digital-economy-and-society-index-desi2017 (Last access: 17 May 2021).

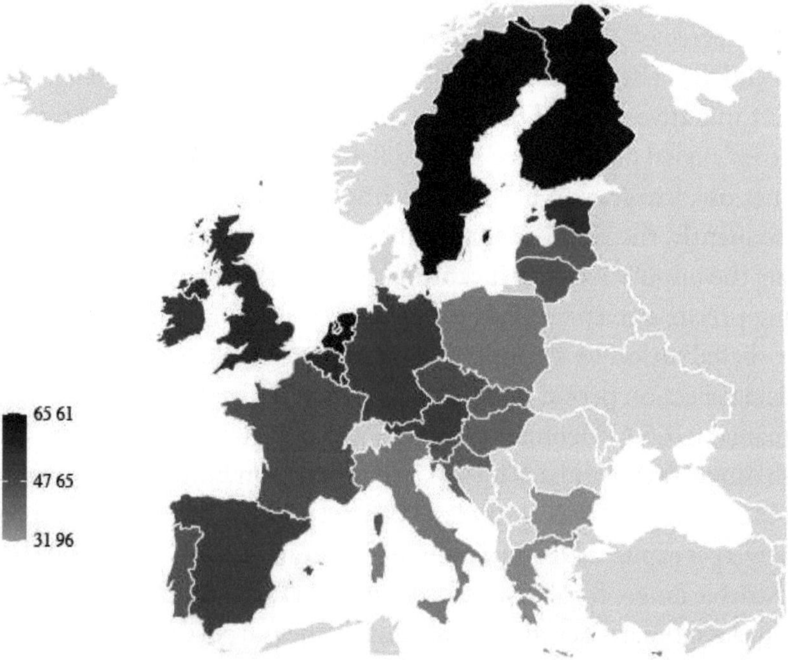

FIGURE 4.3 Distribution of Digital Economy and Society Index (DESI) 2017 across EU countries.

Source: European Commission, Digital Scoreboard via Eurostat empty model allowing an estimation of the amount of variance at the country level. Second, we estimated the fixed effects at the individual level models including a random intercept at the country level. By adding independent and mediating variables stepwise, these models test Hypotheses 1 and 3. The hypotheses involving the contextual level (Hypotheses 2 and 4) are tested by adding contextual information in the fixed effects part, by allowing a random slope for education to vary across countries, and adding a covariance term between the random slope and the random intercept. In these last models, individual-level control variables are included, but only the direct effect of education (without mediation) is estimated. Models are evaluated by means of variations in the explained variance (compared to the variance in the empty model), -2 Log Likelihood and by the likelihood-ratio test (which compares nested models, usually the previous model unless differently specified).

All the analyses are performed with StataMP 16 (StataCorp., 2019); the user-written package polychoric (Kolenikov & Angeles, 2004) was used; the map in Figure 4.3 was produced with R (R Core, 2013) using the packages ggplot2 (Wickham, 2016), mapproj (McIlroy et al., 2018) and rworldmap (South, 2011).

4 • RESULTS

The empty model in Table 4.3 suggests that the country level accounts for 10% of the total variance of e-privacy management (Intraclass Correlation Coefficient = 0.102). Figure 4.4 shows the distribution of the average number of e-privacy management activities for each country.

The first part of the analyses is reported in Table 4.3. Education appears to have a small yet significant and positive effect on managing privacy, as hypothesized, and although the coefficient decreases in size when adding the mediating variables, it holds across models. Taking M1, the increment in e-privacy management activities at each additional year spent in education is small in size (b

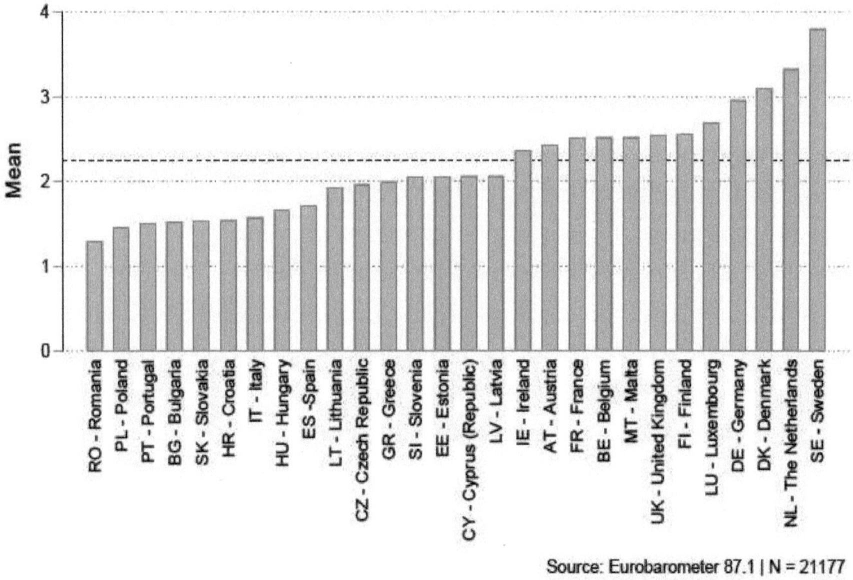

FIGURE 4.4 Average number of e-privacy management activities by country. The dotted line represents the grand-mean across countries.

= 0.07, $p < 0.001$), yet the difference between the lowest (0) and highest (22) amount of years in education reaches about 1.5 activity (out of 8). The models account for variation at the country level, due to composition effects. At the individual level, education and the control variables only explain 3.4% of the variance; the final model (M5), which also includes the mediating variables, explains in total 10.4% of the variance at the individual level.

Hypotheses 1 and 3 are supported, since all the mediating effects play a role in the relationship between education and e-privacy management: frequency of internet use and digital skills, and the reflexive mindset, all display significant and positive coefficients. Digital skills account for the strongest reduction in the effect of education, and explain more variance at the individual level compared to the variance explained by internet use. As for reflexive modernization, albeit own responsibility to tackle climate change shows a significant and positive impact on e-privacy management, it accounts for a small reduction in the direct effect of education. An alternative operationalization of reflexive mindset yields very similar results (see Table C.2 in Appendix C).

The change in standard deviations of the outcome variable at each 1-standard deviation increase in the predictor is obtained by standardizing the b coefficients. This allows to directly compare the strength of the coefficients of different predictors within a model. The standardized coefficients (β) in M5 show that digital skills have a stronger impact ($\beta = 0.21$) on e-privacy management compared to believing that it is one's own responsibility to tackle climate change ($\beta = 0.06$) and internet use ($\beta = 0.11$). All coefficients, anyway, remain significant, positive and substantial.

With respect to the control variables, women are consistently less prone to manage their privacy than men; the negative effects of age and unemployment, as well as the positive effect of being students are also explained by the digital divide, as their coefficients drop in size when adding internet use and/or digital skills to the models. The type of settlement only has limited impact on the tendency to manage privacy.

In order to explore the residual direct effect of education on e-privacy management, we break it down by age groups (see Figure 4.5). Even net of digital skills and internet use, it is mostly among the adult age groups (35–54 years old) that the highest educated (90th percentile) individuals tend to enact significantly more e-privacy protection activities compared to the lowest educated (10th percentile). That is, the relative gain of having pursued higher education on e-privacy-savviness is stronger for those who are in their working age. One possible explanation is that higher educated adults are more likely to work in the tertiary sector, which – compared to manual jobs - may actively require e-privacy management. There is no educational gap among the elderly, for

TABLE 4.3 Multilevel linear regression analyses of e-privacy management on individual characteristics (N=21,177).

Variables Fixed effects	M0 b	M1 b	M2 b	M3 b	M4 b	M5 b
Years in education[a]		0.07***	0.06***	0.04***	0.06***	0.04***
Female		−0.12***	−0.12***	−0.07**	−0.14***	−0.09***
Age categories						
15-24		0.14*	0.08	0.05	0.14*	0.02
25-34		0.01	−0.02	−0.05	0.01	−0.07+
35-44 (Ref.)						
45-54		−0.04	−0.001	0.03	−0.04	0.05
55-64		−0.15***	−0.06	−0.001	−0.14**	0.05
65-74		−0.24***	−0.14*	−0.04	−0.22***	0.02
75+		−0.64***	−0.46***	−0.29***	−0.61***	−0.19*
Student (full time)		0.15*	0.15*	0.09	0.15*	0.10
Unemployed		−0.09+	−0.02	−0.04	−0.09+	0.001
Rural area/village (Ref.)						
Small/middle town		−0.05+	−0.06+	−0.05+	−0.05	−0.06*
Large town		0.04	0.01	−0.00	0.04	−0.01
Internet use[a]			0.44***			0.31***
Digital skills[a]				0.60***		0.52***
Climate change: own responsibility					0.33***	0.29***
Constant	2.19***	2.34***	2.30***	2.27***	2.25***	2.18***
Random effects						
var(Country)	0.37***	0.37***	0.32***	0.29***	0.33***	0.24***
Pseudo R^2country		1.1%	12.9%	22.1%	11.8%	34.5%
var(Individual)	3.23***	3.12***	3.04***	2.95***	3.10***	2.89***
Pseudo R^2individual		3.4%	6.1%	8.8%	4.0%	10.4%
Model fit						
LR chi2		740.7***	591.3***	1214.1***b	134.3***b	1608.6***b
Δdf		12	1	1	1	3
-2 Log Likelihood	85067.5	84326.8	83735.4	83112.7	84192.5	82718.2

[a] Centered variable; [b] Nested in M1;
b = coefficient; var = Variance; LR = Likelihood Ratio
+ $p< 0.10$; * $p < 0.05$; ** $p < 0.01$; *** $p < 0.001$

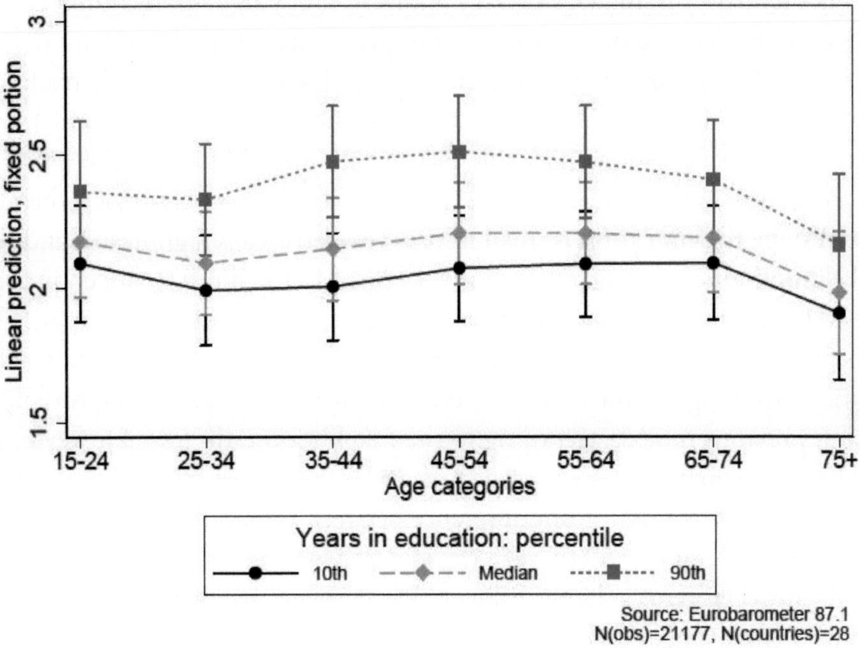

FIGURE 4.5 Predicted values of e-privacy management and 95% confidence intervals by 10th, 50th and 90th percentile of years spent in full-time education and age categories (controlling for all variables included in M5).

whom digital technologies were hardly available during their formation years. For the younger cohorts, an explanation of the lack of the educational gap in e-privacy management may be due to the fact that digital training in school occurs earlier compared to previous cohorts, diminishing the relative gain of each additional year spent in education.

Turning to the country-level models, a country's degree of digitalization has a significant positive effect on e-privacy management , indicating an overall higher tendency to manage privacy online in more digitalized countries; the effect holds also when a country's GDP per capita is added to the model. A large portion of variance (70%) at the country level is explained when adding DESI (M6 in Table 4.4). GDP does not contribute to improve the model, and further inspection shows high collinearity with the DESI index (rho = 0.61).

As concerns the differential effect of education across countries, the random slope for education is small in size, yet statistically significant, and the likelihood-ratio test suggests an improvement in the model (M6, compared to M1): this indicates that the effect of education varies across countries. The coefficient for the cross-level interaction between DESI and years spent in education proves significant at the 90% level, and there is only a small improvement from previous models according to the Likelihood-ratio test. However, by plotting the marginal effects (see Figure 4.6) it can be seen that at lower values of DESI there is a significant difference between those in the 90th percentile of years in education and those in the 10th and median cutoffs. However, this difference disappears at higher values of DESI, suggesting that in countries that underwent stronger digitalization processes the educational gap in e-privacy narrows (supporting Hypothesis 2). These results also lead to reject the expectation that the educational gap in e-privacy would be larger in more digitally advanced countries (Hypothesis 4).

5 • SUMMARY OF FINDINGS AND DISCUSSION

Digital transformations have made privacy a key issue of our time. In this study, we departed from the idea that the level of education – one of the strongest factors of stratification nowadays (cf. Bovens & Wille, 2017) – affects the extent to which individuals protect their privacy online, potentially generating inequalities in the exposure to unwarranted algorithmic processes and cybercrime. Earlier studies yielded mixed findings, and the theoretical links underlying such relationship remained unexplained. To tackle this gap, we tested two

TABLE 4.4 Multilevel linear regression analyses of e-privacy management on individual and country characteristics (N=21,177).

Variable Fixed Effects	M6 b	M7 b	M8 b
Years in education[a]	0.07^{***}	0.07^{***}	0.07^{***}
DESI[a]	0.05^{***}	0.05^{***}	0.05^{***}
GDP per capita[a]		0.002	0.002
DESI*Years in education			−0.001+
Control variables omitted from output			
Constant	2.35^{***}	2.35^{***}	2.35^{***}
Random effects			
var(Years in education)	0.001^{***}	0.001^{***}	0.001^{***}
var(Country)	0.10^{***}	0.10^{***}	0.10^{***}
Pseudo R^2	71.9%	73.0%	73.0%
Covariance Years in education with Country	0.05	0.05	0.04
var(Individual)	3.11^{***}	3.110^{***}	3.11^{***}
Pseudo R^2	3.8%	3.8%	3.8%
Model fit			
LR chi2	70.8^{***b}	1.1	3.0+
Δdf	3	1	1
-2 Log Likelihood	84255.9	84254.4	84251.9

[a] Centered variable; [b] Nested in M1
b = coefficient; var = Variance; LR = Likelihood Ratio
+ p< 0.10; $^{*}p < 0.05$; $^{**}p < 0.01$; $^{***}p < 0.001$

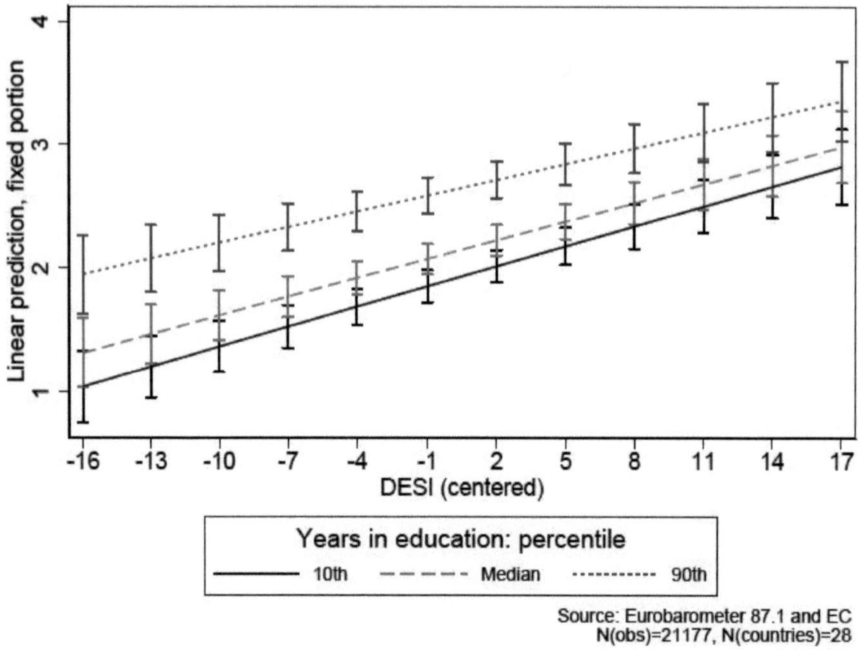

FIGURE 4.6 Predicted values of e-privacy management and 95% confidence intervals by 10th, 50th and 90th percentile of years spent in full-time education and DESI.

potential mechanisms explaining the effect of educational level on e-privacy management, and studied whether this educational gradient varied across more and less digitalized countries.

Our main finding, i.e. the positive and significant association between education and enacted e-privacy protecting activities, aligns with the findings of Y. J. Park and Chung (2017), who found a positive association between the level of education and e-privacy control among US adults. Although the studies by

Park 2011, 2013 did not detect a direct effect of education on e-privacy skills, he found an effect of technical knowledge on e-privacy skills, which is similar to our next set of findings.

Our results indicate that the digital divide theory is highly relevant when it comes to e-privacy management, since both frequency of internet use and digital skills positively affect e-privacy management (cf. Bartsch & Dienlin, 2016; Büchi et al., 2017; Y. J. Park, 2011, 2013) and mediate the effect of education. The lower educated (but also elderly and unemployed) are less equipped to deal with the challenges of advanced digitalization. Due to their lower digital skills and their lower tendency at protecting personal information online, the lower educated result more vulnerable to 'offline' consequences, such as (cyber)crime or unwarranted algorithmic profiling. In light of this finding, the emerging new research on algorithmic skills (Hargittai et al., 2020) and, more generally, on digital inequalities in the algorithmic era (Lutz, 2019), constitute a promising venue for future research.

Theorizing on reflexive modernization only proves partially useful to explain the educational gap in e-privacy management. At the individual level, even though the reflexive mindset positively affects e-privacy management, the evidence of the mediation is weak, and the predictions related to the differences across countries do not hold. Beyond the limitations of the operationalization of the reflexive mindset, there are theoretical aspects to consider. Reflexive modernization theory posits that those who possess more knowledge are better equipped to analyze the risks of modernity itself; in turn, we considered risk awareness as a motivation to engage with privacy management. However, many studies suggested a privacy paradox, which entails a discrepancy between the strong privacy concerns of people and their low tendency to actively protect their privacy online, especially among young people (Acquisti et al., 2015; Blank et al., 2014; Büchi et al., 2017; Hargittai & Marwick, 2016; Kokolakis, 2017; Y. J. Park & Shin, 2020). This paradox may offer an alternative explanation as to why the perception of modern risks does not translate into more e-privacy management even among tech-savvy social segments.

Another reason why the expectations derived from reflexive modernization theory fail to find confirmation here may be the selection of countries. This study focused on countries within the European Union, which – differences notwithstanding – may all be considered to be reflexively modern. Setting aside potential limitations in data collection, expanding to countries with varying levels of internet freedom and online governmental censorship may offer opportunities to investigate the impact of different kinds of digital risks, and the unequal perception thereof, on e-privacy control in a more comprehensive way.

Our findings support the diffusion of innovation theory, and showed that higher and lower educated tend to protect their online privacy to a similar extent in countries where digital processes are widespread, such as Nordic countries (Scandinavia, Netherlands). In Southern/Eastern countries (e.g. Romania, Italy) the educational gap in e-privacy management persists. What remains unclear is which of the many factors constituting the digital readiness of a country drives the narrowing of the educational divide across contexts. Educational systems may play a role, since in countries where schools (and not only universities) are equipped with digital devices and training, digital skills may spread more easily across different social strata. Although in 2019 nearly all EU countries included digital competences at each of the three main educational levels (Bourgeois et al., 2019), differences among countries persist (European Commission, 2019). Future research should focus on this aspect, considering also the accelerating effect that the COVID-19 pandemic has brought to the digitalization processes by, e.g., forcing online education at all school levels.

The diffusion theory also posits that the diffusion process repeats itself at the introduction of each new successful innovation (Rogers, 2003). This means that even in highly digitalized countries, new divides may open up as old ones close. In this study, the measure of e-privacy management refers to rather general behaviors that may have normalized over the years. Recent studies invite to continue to research the topic because, even in digitalized countries, divides in privacy-related behaviors, such as disclosing sensitive information or unknowingly giving up personal data, may occur. For instance, Y. J. Park (2018) found that disclosure of political views on social media – and the consequent ability to engage in online communities - was more frequent among younger and higher-educated men. At the same time, in Chapter 2, findings suggested that, in a highly digitalized country such as the Netherlands, higher educated individuals are more wary of social media. Another study found that younger people and those with higher income and education had a higher likelihood to employ apps for health-related issues: compared to one-to-one exchanges, e.g. emails and texts to the GP, those systems involve the release of personal information to a third party, which may expose users to unwanted consequences (Y. J. Park & Shin, 2020).

This issue of the opening of new divides when new technologies are introduced warns caution in interpreting the finding that education does not affect e-privacy management among younger cohorts. As technologies evolve, so do systems to harvest personal data. Today's youth will probably need advanced digital skills to effectively manage their privacy as tomorrow's adults, it may be that the general notions learnt in school to protect e-privacy in this moment will not suffice in the future, exacerbating divides between those who pursued higher/specialized education an those who did not.

Our study presents some limitations in terms of measurements. First, the measures of e-privacy management refer to a rather general use of the internet and/or to the use of pcs. Although a general measure can work well in general-purpose surveys such as the Eurobarometer, it also does not fully account for the diffusion of mobile devices, which constitute the primary access to internet for many people in lower social strata nowadays. This may lead to an underestimation of the e-privacy management activities among lower-educated people. Moreover, our study is limited to a specific type of e-privacy management linked to cybersecurity. A more encompassing measure of e-privacy management should take into account the communication and social aspects of disclosing personal information on apps and social media. Secondly, the measure of education as the number of years spent in education does not allow to properly distinguish among levels of educational achievements, endangering comparability across countries. Finally, some studies found discrepancies between self-reported and observed digital skills but also, more importantly, showed that these discrepancies depended on socio-demographic characteristics, e.g., gender and income (see review by Litt, 2013b). Systematic bias in this instance may lead to serious flaws in any study tackling the digital divide using self-reported measures, and invites more research to assess and improve the quality of the measurement while maintaining feasibility, especially in general-purpose surveys.

In conclusion, in our study we showed that there are educational gaps in e-privacy among the European general population and that they mostly pass through inequalities in the (skillful) use of internet and not through risk awareness. In addition, however, we also found that a higher level of digitalization in a country smoothens educational differences in e-privacy management. Our findings indicate that effective policies to tackle the reproduction of inequalities in the digital environment should focus on strengthening citizens' digital competences (Büchi et al., 2017). This should not be left to individual initiative and resources, but be part of a larger collective effort, so that everyone can profit from, and possibly contribute to, the digital developments of a country.

4 THE CLOSING EDUCATIONAL GAP IN E-PRIVACY MANAGEMENT

Switch on the Big Brother!

Investigating the educational gradients in acceptance of online and public areas surveillance among European citizens

ABSTRACT

In this study we investigate whether, and why, individuals express different levels of acceptance of surveillance depending on their educational level, and whether this relationship varies with the level of digitalization and globalization expansion of their country. Additionally, we ask whether the type of surveillance (online surveillance vs cameras in public areas) conditions these differences. We build on two theoretical frameworks, one concerned with the resurgence of authoritarian values via the cultural backlash, and the other one explaining how different people analyse manufactured risks differently due to processes of reflexive modernization. In order to test the hypotheses, we employ data from the latest wave of the European Values Study (EVS) and implement multilevel multivariate regression models. Findings indicate that the lower educated individuals are more prone to accept online surveillance, due to their stronger authoritarianism and weaker reflexive mindset; however, there is no educational gradient in acceptance of video surveillance in public areas. Additionally, the countries' levels of digitalization and globalization expansion do not condition the educational gradient in acceptance of surveillance.

This chapter, with slight differences, has appeared in print. See Maineri, A., Achterberg, P., & Luijkx, R. (2022). Switch on the Big Brother! Investigating the educational gradients in acceptance of online and public areas surveillance among European citizens. *European Societies.* https://doi.org/10.1080/14616696.2022.2043412. Online Supplementary materials and replication materials are available on OSF (https://doi.org/10.17605/OSF.IO/M82KW).

1 • INTRODUCTION

Alongside the development of sophisticated surveillance technologies, questions about the opportunity of their adoption emerge. The deployment of surveillance is presented by the monitoring institutions as a shield from physical risks, deemed to increase safety and the perception thereof (Trüdinger & Steckermeier, 2017; van Heek et al., 2017). The technological component is justified by the increase in efficiency and accountability, and by the reduction of bias (Brayne, 2017). However, surveillance generates risks for the monitored subjects (Wester & Giesecke, 2019) and is privacy-intrusive (van Heek et al., 2017, p. 80), as individuals may lose control over what is known about them and by whom, or feel the need to self-censor themselves to comply with the monitoring activities (Degli Esposti & Santiago Gómez, 2015). Moreover, technology-driven surveillance tools are not unbiased: the collection, processing, and recombination of data reinforces existing inequalities by disproportionately targeting already-vulnerable social groups, e.g. people with a low SES, or minorities (Brayne, 2017; Lutz, 2019; Mann & Matzner, 2019) (see also Chapter 1).

Nowadays, lower educated individuals can be considered vulnerable. Not only a high level of education is more rewarded in the de-industralized labour markets of advanced economies (Bonoli, 2007), but also in daily lives education enables people to navigate the complexity of the digital word more efficiently, as research on the digital divide has showed (for extensive reviews, see Lutz, 2019; Scheerder et al., 2017). Lower educated individuals are also more exposed to cybercrime and social sorting by lagging behind with online privacy protection, as shown in Chapter 4 and by Y. J. Park and Chung (2017).

A paradox emerges, since - despite their stronger exposure to harmful digital surveillance processes - lower educated individuals have been found to accept surveillance to a larger extent (Trüdinger & Steckermeier, 2017; van den Broek et al., 2017). In these studies, education was only used as a control variable, underscoring the importance of exploring the mechanisms underlying this relationship. We propose two distinct explanations. While both theories predict higher acceptance of surveillance among lower educated individuals, they stress different sides of the tension between security and privacy risks unfolding within surveillance. The first explanation emphasizes the security side and draws on the cultural backlash thesis (Norris & Inglehart, 2019), according to which lower educated individuals demand more surveillance because of their stronger authoritarian attitudes, ignited by the expansion of leftist-progressive policies and globalization. The second explanation focuses on privacy risks and, drawing on the reflexive modernization theory (Beck, 1992), sees the larger tolerance of surveillance among the lower educated individuals as rooted in a lack of critical mindset, which hinders their ability to recognize the risks entailed by a modern institution, i.e. government surveillance (cf. Chapter 1). Albeit the two theories are not completely alternative to each other when it comes to differences between individuals, the derived expectations are conditional on the type of surveillance under scrutiny and on the national context.

The research problem is therefore addressed at three levels. First, we look at differences among individuals. Lower educated individuals tend to hold authoritarian values (Stubager, 2008; van de Werfhorst & de Graaf, 2004), and lack awareness over the functioning of institutions, essential to recognize 'modern' risks (Makarovs & Achterberg, 2017): both mechanisms could explain their stronger support for surveillance. Second, individual-level mechanisms may be conditional upon the national context: while the cultural backlash hypothesis emphasizes how rapid social changes trigger a security demand among vulnerable social groups, reflexive modernization illustrates the knowledge gaps created by the diffusion of ICTs. Third, the type of surveillance matters for its acceptance among citizens (van den Broek et al., 2017), with surveillance technologies in public areas more widely accepted than in private areas (Degli Esposti & Santiago Gómez, 2015; van Heek et al., 2017). A privacy intrusion caused by surveillance is perceived as more problematic in online settings than in the public space; yet, the authoritarian appeal is likely to apply to both types of surveillance. In this study we gradually address the following questions:

RQ1 *To what extent and why is there an educational gradient in acceptance of surveillance?*
RQ2 *Is the educational gradient in acceptance of surveillance stronger in online settings compared to public areas?*
RQ3 *To what extent is the educational gradient in acceptance of surveillance conditioned by the degree of digitalization and globalization expansion in a country?*

Whereas previous studies on the acceptance of surveillance adopted a qualitative approach (Degli Esposti & Santiago Gómez, 2015), focused on one country (Trüdinger & Steckermeier, 2017; Wester & Giesecke, 2019) or relied on convenience samples (van Heek et al., 2017), we add to the literature by using data from the European Values Study (EVS) 2017, which has been collected among representative samples of individuals from over 30 European countries.

2 • EXPLAINING ACCEPTANCE OF SURVEILLANCE

Types of surveillance

Contemporary surveillance systems (or 'dataveillance', cf. van Dijck, 2014) often rely not only on the collection of information about individuals and their activities, but they also gather contextual information carrying great disclosure potential when re-combined. Facial recognition technologies, for instance, allow to follow single individuals moving across different places. Despite the high level of privacy-intrusiveness, surveillance, whether it is a monitoring of public areas or of internet communications, can be promoted by the institutions a security measure, as seen with the justification provided for the collection of internet data within the PRISM program unveiled by Edward Snowden (Lyon, 2014).

However, the type of surveillance has implications for its acceptance among citizens (Degli Esposti & Santiago Gómez, 2015; van den Broek et al., 2017; Wester & Giesecke, 2019); for instance, people are more inclined to accept cameras in public locations than in private areas (Degli Esposti & Santiago Gómez, 2015). In public spaces, it is implied that one's behaviour is already visible to others, and there is a large consensus over what constitutes appropriate behaviour, in compliance with the law but also with social norms (Hatuka & Toch, 2017). Surveillance in public areas shields from risks which are socially acknowledged as such: deviance is possible, but it can be punished. Surveillance of communications exchanged online, instead, makes monitored subjects feel personally targeted, since the definition of what constitutes a risk online is less clearcut (Degli Esposti & Santiago Gómez, 2015). Whilst it is relatively clear what behaviour, if monitored in a public area, would produce a reaction from authorities, this is not straightforward when it comes to online behaviour. Therefore, online surveillance may be perceived as more privacy-intrusive than surveillance in public areas.

The educational gradient

SECURITY DEMAND AND CULTURAL BACKLASH

Surveillance as a way of protecting security is instrumental to the desire for 'law and order', which has been resurging in recent decades. The cultural revolution of the 1960s-1970s, which led to a more progressive political and cultural climate, created the space for the revival of conservative values starting from the 1980s (de Koster et al., 2008). The post-materialist value shift proposed by Inglehart (1977, 1981), has emphasized the libertarian side of this new cultural climate: accordingly, due to the increased economic security after the second world war, people's priorities shifted from material issues (e.g. physical safety) to immaterial, or post-material, issues (e.g. self-expression). However, this perspective has failed to grasp the emergence of the neo-conservatist cultural climate which, combined with other factors, explains the success of extreme right-wing parties in Europe in the last decades (Ignazi, 1992). According to Flanagan (1987), the loss of salience of economic issues has led to a non-materialist value change, in two different directions: libertarian (overlapping with Inglehart's understanding of post-materialism), and authoritarian. The authoritarian value pattern prioritizes non-economic issues, among which 'law and order' (Flanagan, 1987).

Education stands out among the individual characteristics affecting the authoritarian/libertarian value dimension (Flanagan, 1987). Education is thought to affect political attitudes via socialization (Stubager, 2008; van de Werfhorst & de Graaf, 2004): students internalize values during their school years, and when in higher education a 'view is promoted that social reality is an ongoing human product in which individual action can make a difference' (van de Werfhorst & de Graaf, 2004, p. 215). As a result, higher educated individuals are generally more tolerant and libertarian compared to lower educated ones, who instead display authoritarian traits (Stubager, 2008; van de Werfhorst & de Graaf, 2004).

After dismissing authoritarian values as a matter of old politics, Inglehart has revised his position, and proposed the concept of the cultural backlash (Norris & Inglehart, 2019) to indicate the resurgence of authoritarian values among the vulnerable strata of the population (e.g., elderly, less educated individuals), as a reaction to the progressive-leftist policies threatening family values, and to the rapid growth of social diversity brought along by globalization (Norris & Inglehart, 2019). Accordingly, the cultural backlash explains the support for populist authoritarian movements witnessed in recent decades in Europe and the US (Norris & Inglehart, 2019). Hence, the relationship between education and acceptance of surveillance as an authoritarian reflex should be stronger in countries which underwent rapid social changes.

REFLEXIVE MODERNIZATION AND KNOWLEDGE GAPS

Surveillance creates uncertainty for the monitored subjects, as their increased safety may come at the expenses of individual privacy (Degli Esposti & Santiago Gómez, 2015; Trüdinger & Steckermeier, 2017). Arguably, this aligns with the idea of manufactured uncertainties (Price & Peterson, 2016), key to Beck's risk society thesis (Beck, 1992): in contemporary societies, individuals must co-exist with the threats posed by modern technological advances (Beck, 1992).

Central to risk societies is reflexivity (Beck, 1992), i.e. a critical questioning of modernity itself and of its achievements (Knight & Warland, 2005) coupled with a progressive distancing from a dogmatic interpretation of knowledge (De Keere, 2010). As a process of constant revision of modernity, reflexivity enables skepticism towards 'the notion that secular understandings of the world lead to a safer and more rewarding existence for humans' (Knight & Warland, 2005, p.257). In reflexive modern societies, technological and scientific advances are questioned (Price & Peterson, 2016). For instance, individuals in countries with a higher rate of tertiary education enrolments and internet access displayed higher distrust in science (Price & Peterson, 2016). This critical attitude towards progress enables awareness over the manufactured risks of modernity.

The inclusion in this process of revision of modernity is, however, stratified, as manifest in the 'new social divisions between the "information rich" and "information poor"' (Elliott, 2002, p. 304), with the latter lacking resources to acknowledge the existence of modern risks. Empirical evidence on the role of education in this is mixed (cf. Chapter 1). When taking the lack of confidence in science as an expression of a critical attitude over modern institutions at the individual level, higher educated individuals were found to be more trusting of science rather than less (Achterberg et al., 2017; De Keere, 2010; Price &

Peterson, 2016). Others found, however, that higher educated individuals in reflexively modern societies developed a heightened knowledge and awareness about the functioning of modern institutions rather than a generalized skeptical stance towards them. For instance, a high level of education in reflexively modern societies is negatively associated with the likelihood of getting the seasonal flu shot (Makarovs & Achterberg, 2017), explained by the authors as a critical reaction to scientific progress. A study found that higher educated individuals in the U.S. were not generally more skeptical about science, but that those who were were also more prone to translate their negative views into a lack of support for embryonic stem cell research compared to lower educated individuals (Nisbet & Markowitz, 2014). Hakhverdian and Mayne (2012) found that the level of corruption of a country has detrimental effects on institutional trust only among the higher educated individuals, suggesting that they are better able to critically evaluate the functioning of institutions.

The pre-requisite for the in-depth knowledge of modern institutions is the availability of information, fuelled by the diffusion of ICTs. Albeit vast amount of information is available at little cost to a broader swath of individuals, it has become increasingly difficult to navigate the multiplicity of (often contradictory) sources. Therefore, rather than equalizing knowledge evenly, the diffusion of ICTs has created knowledge gaps, i.e. the differential growth in knowledge between the higher and lower social strata due to the easier and faster uptake of information of the former (Bonfadelli, 2002). This mechanism has been extensively found in the literature (for a review, see Lind & Boomgaarden, 2019). Therefore, the spread of ICTs in a country may facilitate the concentration of awareness about surveillance technology-related risks among the higher educated individuals.

Summary and hypotheses

The outlined theoretical framework, conceptually summarized in Figure 5.1,

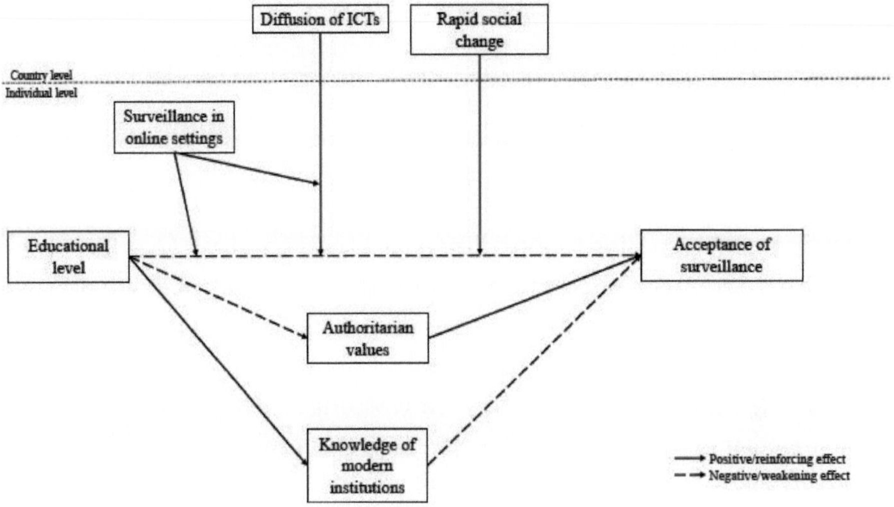

FIGURE 5.1 Conceptual model

underscores the importance of education to understand attitudes towards surveillance, and provides some tentative explanations over the negative educational gradient in acceptance of surveillance (AoS) which was found in previous studies yet left unexplained (Trüdinger & Steckermeier, 2017; van den Broek et al., 2017). In this section, we formalize the mechanisms and formulate hypotheses derived from the two theoretical strands (i.e. cultural backlash and reflexive modernization).

Authoritarian values feed AoS in the name of increased safety (de Koster et al., 2008). Since the lower educated respondents are more sensitive to the safety appeal of surveillance due to socialization processes (Stubager, 2008; van de Werfhorst & de Graaf, 2004), we argue that they are more likely to accept surveillance regardless of the potential level of privacy intrusiveness. Hence, we expect that

HP1 The negative association between education and AoS is equally strong for online surveillance and for public areas surveillance.
HP2 Acceptance of both public areas and online surveillance is higher among the lower educated individuals because of their stronger authoritarian attitudes.

Additionally, according to the cultural backlash hypothesis (Norris & Inglehart, 2019), the authoritarian reflex among the lower educated individuals is accelerated by social change, hence we expect that:

HP3 The educational gradient in AoS is stronger, irrespective of the type of surveillance, in countries which recently underwent rapid globalization expansion.

In contrast, the reflexive modernization perspective (Beck, 1992) suggests that higher educated individuals, because of their knowledge of the functioning of modern institutions, would be better able to analyse the risks, as in gains and drawbacks associated with surveillance (Elliott, 2002; Hakhverdian & Mayne, 2012; Nisbet & Markowitz, 2014; Price & Peterson, 2016), leading to lower acceptance overall but also to a clearer differentiation between the types of surveillance. Due to the sensitivity of the information exchanged and to the differences in exposure and visibility associated with the two types of surveillance (cf. Hatuka & Toch, 2017), we suspect that privacy risks are perceived more distinctly for online surveillance than for surveillance in public areas. Hence, we expect the heightened knowledge over modern institutions to reduce the strength of the relationship between education and AoS, only in online settings. The following hypotheses are formulated:

HP4 The negative association between education and AoS is stronger for online surveillance than for public areas surveillance.

HP5 AoS online is higher among lower educated individuals because of their lower knowledge of modern institutions.

In countries where digitalization processes are more widespread, the larger accessibility to information fosters awareness over the functioning of institutions. However, following the knowledge gap hypothesis (Bonfadelli, 2002), this occurs to a larger extent among higher educated individuals. Accordingly, we expect a steeper educational gradient in the acknowledgment of privacy risks in online surveillance where there is a larger availability of ICTs, flowing into the following hypothesis:

HP6 The negative association between education and AoS in online settings is stronger in more digitalized countries.

3 • DATA AND METHOD

To study the impact of education on AoS at the individual level, we use data from the Integrated Dataset of the EVS, which collected data from representative samples of over 55,000 individuals in 34 countries[1] between 2017 and 2020 (European Values Study, 2020). The main mode of data collection is face-to-face, but six countries implemented a mixed-mode design (see Luijkx et al., 2021).

To measure globalization expansion at the country-level, the study employs the KOF Globalization Index (Dreher, 2006; Gygli et al., 2019), published by the KOF Swiss Economic Institute. This index has been designed to cover economic, political, social and cultural factors related to globalization (Dreher,

[1] See the list of countries in the Appendix D.

2006). The revised dataset comprising data from 1970 until 2020 in all the countries included in EVS is used (for more details, see Gygli et al., 2019).[2]

Data on the level of digitalization of the country was collected by the World Economic Forum in the form of the Network Readiness Index (NRI).[3] The NRI data from 2016 contain information for all the countries in the EVS, except Belarus, which is excluded from the study.

Variables

To measure *AoS*, respondents expressed opinions on whether the government should have the right to:

- Keep people under video surveillance in public areas;
- Monitor all e-mails and any other information exchanged on the Internet.

Answer categories ranged from 1 (Definitely should have the right) to 4 (Definitely should not have the right); the coding has been reversed so that high values indicate high acceptance. Figure 5.2 reports the relative distribution of the two variables, which have a sizeable correlation (Pearson's $\rho = 0.45$, $p < 0.001$).

The main independent variable is the *educational attainment* of the respondent, measured by the European Survey ISCED (ES-ISCED) classification, an adaptation of the ISCED official classification designed by Schneider (2009). The ES-ISCED scale, devised as a metric variable, ranges from 0 (No formal or less than primary education) to 7 (Master's and higher level), excluding 84

[2] The dataset is available at http://www.kof.ethz.ch/globalisation/.

[3] Now collected and published by the Portulans Institute; see https://networkreadinessindex.org.

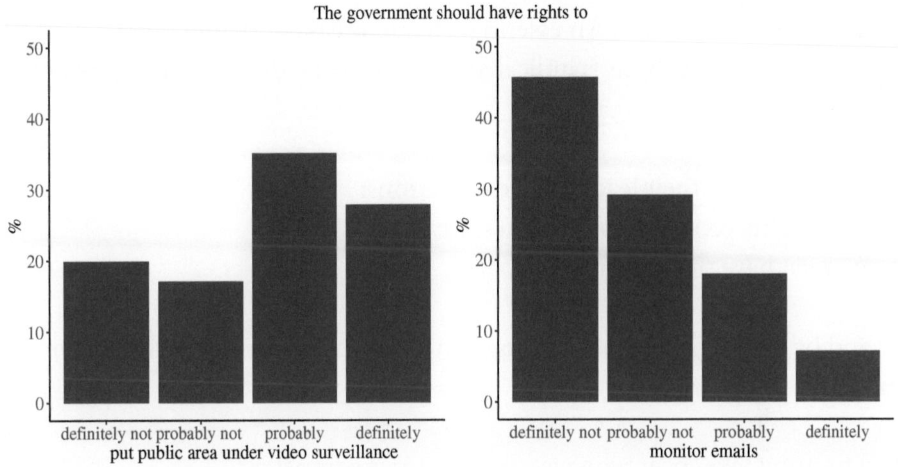

FIGURE 5.2 Distribution of the dependent variables (N = 48,047). Source: EVS (2020), own calculations.

cases with 'other' educational level (see Appendix D for the distribution of educational attainment across countries).

Political authoritarianism is measured by an item from the democracy-autocracy preference scale, asking whether 'Having a strong leader who does not have to bother with parliament and elections' is a very good, fairly good, fairly bad or very bad way of governing this country; after reversing the scale, higher scores indicate stronger preference for a strong leader. The average scores vary across countries, with the lowest support for a strong leader found in Norway, the highest in Georgia (see Appendix D).

The *awareness over the functioning of institutions* is measured by institutional knowledge (cf. Achterberg et al., 2017). Following Wegscheider and Stark (2020), we use a battery measuring the essential characteristics of democracy. The response categories range from 1 (Not at all an essential characteristic of democracy) to 10 (An essential characteristic of democracy).[4] A good knowledge of institutions is indicated by recognizing the following as essential characteristics of democracy:

- People choose their leaders in free elections;
- Civil rights protect people from state oppression;
- Women have the same rights as men;

and by recognizing the following as non-essential characteristics of democracy:

- The army takes over when government is incompetent;
- Religious authorities ultimately interpret the laws;
- People obey their rulers.

Respondents who answered 'Don't know' were also recoded into the low-knowledge category due to the nature of the questions.[5] After ascertaining the clustering of the items with a factor analysis (see Appendix D), we added scores related to knowledge of democracy and the reversed scores related to the knowledge of authoritarian regimes, normalized them in a 0-1 range and

[4] In the self-administered version in Denmark and the Netherlands, category 0 (It is against democracy) was visible for respondents, instead of only coded if spontaneous. For the purposes of this analysis, the 10- or 11-categories versions are not differentiated.

[5] Their answers were recoded to 0 in the three items indicating knowledge of democratic principle, and to 10 in the three items characterizing authoritarian regimes.

subsequently multiplied them.[6] The resulting scores range between 0.00 and 1.00 (see Appendix D for the distribution of the average scores by countries).

Control variables include age, sex (recoded as female) and the mode of data collection to rule out potential mode effects. Table D.2 (see Appendix D) reports the descriptive statistics of the individual-level variables.

The KOF-GI is used to measure *globalization expansion*. To capture the change in globalization rather than its absolute value and better represent the theoretical mechanism, we subtracted the KOF-GI of 2017 from the one of 2007, and labelled it ΔKOF-GI: the higher the score, the more globalization has expanded in that country in the 10-year span. The distribution of ΔKOF-GI scores across countries is presented in Figure 5.3a, ranging from -2.02 (Iceland) to 12.03 (Georgia), with a mean of 3.44 and a standard deviation of 3.43. Due to its skewness, the distribution was normalized using a Box Cox transformation (after shifting the values above 0, and selecting 0.5 as the optimal λ value) and standardized. The resulting score varies between -2.9 and 2.2.

For the country's *level of digitalization*, we use the Network Readiness Index (NRI)(Baller et al., 2016). The NRI, based on 53 indicators, is designed to evaluate how countries leverage digital transformations, and ranges between 3.6 and 5.9 (mean = 4.8, sd = 0.68; see Figure 5.3a). Unsurprisingly, Northern European countries display higher levels of digitalization compared to Southern and Eastern European countries.

Analytical strategy

To test the hypotheses, we employ multivariate multilevel multiple linear regres-

[6] Wegscheider and Stark (2020) justified the choice of multiplying by explaining that 'low knowledge of authoritarian regime principles as non-essential characteristics of democracy cannot be compensated by high knowledge about democratic ones' (Wegscheider & Stark, 2020, footnote 4) and that robustness checks displayed similar results with an additive index.

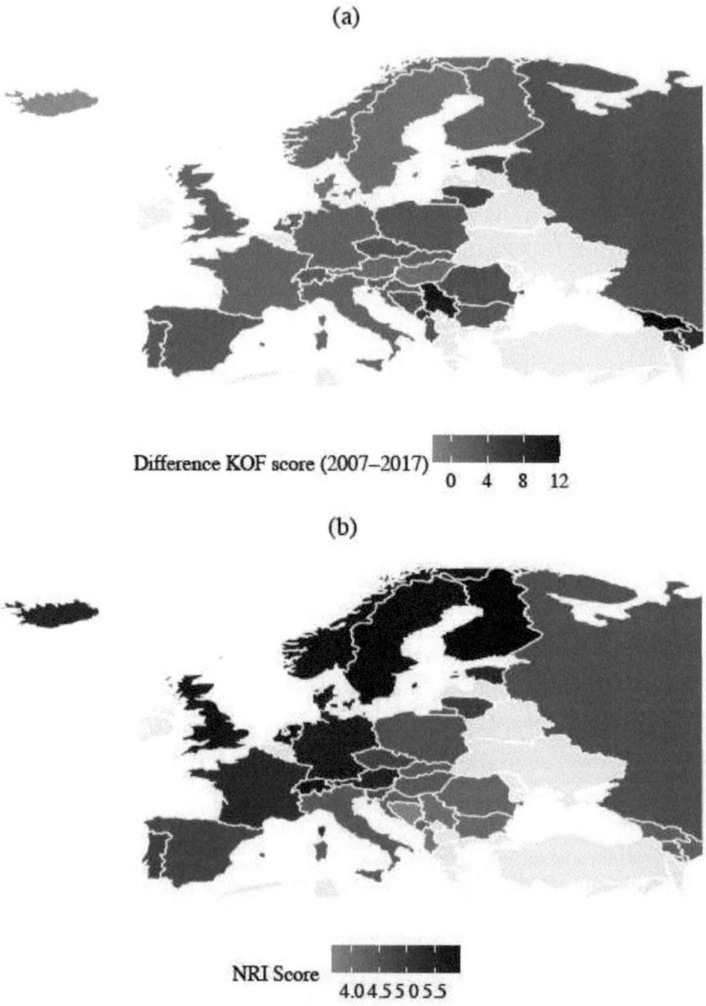

FIGURE 5.3 (a) Distribution of the change in KOF Globalization Index between 2007 and 2017. Source: KOF Globalization Index (1970-2020), own elaboration (b) Distribution of the NRI across countries. Source: World Economic Forum (2016)

sion models. Multiple regression models allow to control for several independent variables simultaneously. The multilevel feature enables to disentangle the variation within and between countries, and to test the interactions between individual- and contextual-level factors. Finally, the models are multivariate because the equations are estimated simultaneously on two highly correlated dependent variables. Through multivariate models it is possible to directly test the equality of certain coefficients or variance components (Hox, 2002). The multivariate multilevel set-up requires the specification of three levels (Snijders & Bosker, 1999): (a) the dependent variables, here specified by two dummy variables, public and online, (b) individuals and (c) countries. All independent variables, and the constant, are multiplied by the DV dummy variables (Snijders & Bosker, 1999), resulting in the estimation of two constants and two sets of fixed effects, one for each dependent variable. The interpretation of the coefficients resembles that of univariate models. In comparison to a classic multilevel model, there are additional variance components, e.g. the correlation between the random effects of the two dependent variables.[7]

We built the models incrementally: after estimating an empty model to assess the variance allocated to each level, models testing the effect of the independent variables were estimated, adding mediators step-wise to better control their impact on the direct effect of education. Subsequently, the random slopes for education were added, to assess changes in the correlation between education and AoS across countries; finally, country-level predictors, interacted with individual educational attainment, were added. For the purposes of the analyses, all independent variables but dichotomous variables have been centred around the grand mean. Analyses are performed on R (R Core Team, 2021) using several packages; the multilevel multivariate models are obtained with the package nlme (Pinheiro et al., 2021).

[7] The estimation of the correlations follows the procedure of Snijders and Bosker (1999) via R syntax at https://www.stats.ox.ac.uk/~snijders/ch16.r.

The analytic sample includes 48,047 individuals nested in 33 countries, after deleting 12.5% of cases due to missing values (see Table D.1 in Appendix D). For preference for a strong leader and institutional knowledge, descriptive analyses showed that cases with missing values (respectively, 6.4% and 2.4%) tend to display lower AoS. However, alternative models (see Online Supplementary Materials) showed that retaining these cases in the analytical sample by imputing values (either via mean imputation or FIML) does not lead to a substantial change in the results compared to the models with listwise deletion, which are hence presented in the manuscript.

4 • RESULTS

At a descriptive level (see Figure 5.4), surveillance in public areas is consistently more accepted than online. This is confirmed by the intercepts of the two variables in the regression models (see Table 5.1, M0), showing an average AoS of 2.66 ($p < 0.001$) for video surveillance in public areas, and 1.84 ($p < 0.001$) in online settings. There is variation across countries, with 11% of the variance of AoS in public areas, and 6% of the variance of AoS online, attributed to the country-level.

As concerns individual-level models, results are displayed in M1 to M4 in Table 5.1. In M1, a negative association between education and AoS is displayed only in online settings. The coefficient is small: at each additional attained level of education, acceptance of online surveillance drops by 0.04 ($p < 0.001$), i.e. the relative difference in AoS online between the lowest and the highest educated individuals is 0.28 on a 4-point scale (see Figure D.4 in Appendix D for the predicted values of AoS by education). Regarding AoS in public areas, the association with education is not statistically significant. The coefficients of education on the two types of AoS are statistically different (Wald chi2=163.6, Δdf=1, $p < 0.001$). These findings lead to reject the expectation of equal strength in the educational gradient regardless of the

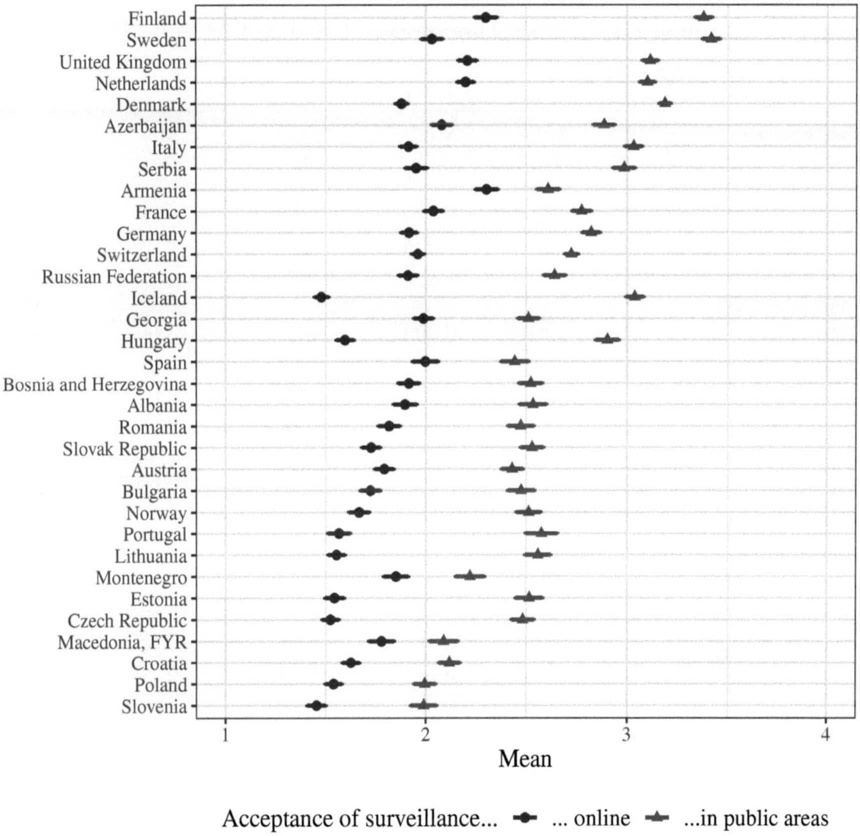

FIGURE 5.4 Average acceptance of surveillance by country with 95% Confidence intervals. Source: EVS (2020). Own calculations.

type of surveillance formalized in HP1, and to accept HP4, predicting the educational gradient to be stronger for online surveillance.

TABLE 5.1 Multivariate multilevel linear regression of AoS on individual characteristics (N=48,047 in 33 countries).

Fixed effects Predictor	M0 b	M1 b	M2 b	M3 b	M4 b
Type surveillance: public x	2.66^{***}	2.65^{***}	2.65^{***}	2.65^{***}	2.65^{***}
Education		0.001	0.004	0.01^{*}	0.01^{*}
Preference for strong leader			0.03^{***}		0.02^{***}
Inst. knowledge				-0.17^{***}	-0.14^{***}
Age		0.003^{***}	0.003^{***}	0.003^{***}	0.004^{***}
Female		−0.01	−0.01	−0.01	−0.01
Mode: CAWI		0.09^{***}	0.09^{***}	0.09^{***}	0.09^{***}
Mode: Mail		0.07^{*}	0.07^{*}	0.06^{*}	0.06^{*}
Type surveillance: online x	1.84^{***}	1.83^{***}	1.83^{***}	1.83^{***}	1.83^{***}
Education		-0.04^{***}	-0.03^{***}	-0.02^{***}	-0.02^{***}
Preference for strong leader			0.09^{***}		0.05^{***}
Inst. knowledge				-0.59^{***}	-0.53^{***}
Age		0.01^{***}	0.01^{***}	0.01^{***}	0.01^{***}
Female		−0.001	0.001	−0.01	−0.01
Mode: CAWI		0.06^{*}	0.06^{*}	0.08^{***}	0.08^{***}
Mode: Mail		0.11^{***}	0.11^{***}	0.10^{***}	0.10^{***}
Random effects					
Var(Country | public)	0.14	0.13	0.13	0.13	0.13
Var(Country | online)	0.06	0.06	0.05	0.05	0.05
Cor(Country | public, Country | online)	0.54	0.52	0.57	0.56	0.59
Var(Individual | public)	1.04	1.04	1.04	1.04	1.04
Var(Individual | online)	0.86	0.84	0.84	0.83	0.82
Cor(Individual | public, Individual | online)	0.44	0.44	0.44	0.44	0.44

b = coefficient; Var = Variance; Cor=Correlation
$^{*}p < 0.05$; $^{**}p < 0.01$; $^{***}p < 0.001$

Both preference for a strong leader and institutional knowledge have significant and, respectively, positive and negative, associations with AoS. The magnitude of the effects is stronger for acceptance of online surveillance compared to acceptance of video surveillance in public areas: taking M4, when comparing the most knowledgeable individuals to the least knowledgeable ones, AoS drops by 0.14 when in public areas and by 0.53 when in online settings.[8]

Turning to the mediating effects, preference for a strong leader and low institutional knowledge among lower educated individuals[9] partially explain the educational gradient in AoS online. The coefficient indicating the educational gradient in AoS ($b = -0.04, p < 0.001$, cf. M1) shrinks in size when adding the mediators. Institutional knowledge accounts for a stronger reduction in the direct association between education and AoS in online settings ($b = -0.02, p < 0.001$, cf. M3) compared to the preference for a strong leader ($b = -0.03, p < 0.001$, cf. M2). For surveillance in public areas, the coefficient for education remains small in size, but turns positive and significant ($b = 0.01, p < 0.05$, cf. M4). This leads to a partial support of HP2, since authoritarian attitudes mediate the association between education and AoS when online but not when in public areas. However HP5, predicting the educational gradient in AoS online - and not that in AoS in public areas - to be explained by institutional knowledge, is supported.

The explanatory power of the models is weak. When including all individual-level independent variables (M4), only 4% of the variance of AoS in online settings and $< 1\%$ of the variance of AoS in public areas is explained. The

[8] Concerning control variables, AoS is positively associated with age, whereas the association with sex is not significant. Respondent in the self-administered mode show slightly higher AoS than respondents interviewed face-to-face.

[9] Preference for a strong leader has a negative correlation with education ($\rho = -0.15, p < 0.001$) whereas institutional knowledge has a positive correlation with education ($\rho = 0.21$, $p < 0.001$), in line with the theoretical expectations.

last row of Table 5.1 reports the residual correlation between the two types of surveillance, which remains stable, indicating that the individual-level characteristics hereby considered do not account for the correlation between the two types of AoS.

As for country-level models, the results are reported in M5 to M7 in Table 5.2. The random slopes for education, Var(Education | public/online), are negligible in size (0.001, cf. M5). However, a conditional likelihood ratio test ($LRchi2 = 144.8, \Delta\, df = 7, p < .0001$) indicates an improvement in the random slope model (M5) compared to the random intercept model including the same fixed effects (M1), indicating variation across countries in the educational gradient in AoS (see Figure D.5 in Appendix D).

For both types of surveillance, the association between education and AoS does not vary with ΔKOF-GI (see Figure 5.5), as the coefficients of the interactions are not significant (respectively, $b = 0.01, p > 0.05$, and $b = 0.002$, $p > 0.05$; cf. M6), and little additional variance is explained at the country-level. Given that the educational gradient in AoS is not larger in countries that underwent rapid globalization expansion, HP3 is rejected.

The cross-level interaction between the level of digitalization and education on AoS in online settings is also not significant ($b = 0.01, p > 0.05$; cf. M7 and Figure 5.6). Since the negative educational gradient in AoS online is not steeper in more digitalized countries, HP6 is rejected. Interestingly, AoS in public areas is higher in more digitalized countries and explains 20 percentage point of country-level variance in addition to M5. As for AoS in online settings, however, there is no variation by levels of NRI, and little additional variance is explained.

5 • DISCUSSION

The definition of the governments' rights to monitor citizens using digital surveillance tools constitutes a key challenge for policymaking nowadays. Despite the safety narrative promoted by monitoring institutions, surveillance generates risks for the monitored subjects, especially in terms of privacy inva-

TABLE 5.2 Multivariate multilevel linear regression of AoS on individual and country characteristics (N=48,047 in 33 countries).

Fixed effects Predictor	M5 b	M6 b	M7 b
Type surveillance: public x	2.65***	2.64***	2.65***
Edu.	0.01	0.01	0.00
Age	0.003***	0.003***	0.003***
Female	−0.01	−0.01	−0.01
Mode: CAWI	0.09***	0.09***	0.09***
Mode: Mail	0.06	0.06	0.06
ΔKOF-GI		−0.11	
Edu. x ΔKOF-GI		0.01	
NRI			0.25*
Edu. x NRI			0.001
Type surveillance: online x	1.83***	1.84***	1.83***
Edu.	−0.04***	−0.04***	−0.04***
Age	0.01***	0.01***	0.01***
Female	0.000	0.000	0.000
Mode: CAWI	0.06*	0.06*	0.06*
Mode: Mail	0.11***	0.12***	0.11***
ΔKOF-GI		0.08	
Edu. x ΔKOF-GI		0.002	
NRI			0.04
Edu. x NRI			0.01

Table continues on next page.

Random Effects Predictor	M5 b	M6 b	M7 b
Var(Country | public)	0.13	0.12	0.10
Var(Country | online)	0.06	0.05	0.06
Cor(Country | public, Country | online)	0.51	0.67	0.52
Var(Edu. | public)	0.001	0.001	0.001
Var(Edu. | online)	0.00	0.00	0.00
Cor(Edu. | public, Edu. | online)	0.71	0.72	0.72
Cor(Country | public, Edu. | public)	−0.41	−0.38	−0.46
Cor(Country | public, Edu. | online)	−0.26	−0.33	−0.26
Cor(Country | online, Edu. | public)	−0.16	−0.15	−0.26
Cor(Country | online, Edu. | online)	0.04	0.03	0.02
Var(Individual | public)	1.03	1.03	1.03
Var(Individual | online)	0.84	0.84	0.84
Cor(Individual | public, Individual | online)	0.44	0.44	0.44

b = coefficient; sig = Significance; Var = Variance; Cor=Correlation; Edu. = Education
$^{*}p < 0.05$; $^{**}p < 0.01$; $^{***}p < 0.001$

sion (cf. Chapter 1). Inspired by earlier studies suggesting a stronger acceptance of surveillance among lower educated individuals (Trüdinger & Steckermeier, 2017; van den Broek et al., 2017), we investigated the underlying mechanisms of this relationship, as well as their conditionality upon national contexts and types of surveillance. We aimed at understanding whether, and why, social groups potentially more exposed to the negative consequences of extensive surveillance may also be more willing to grant the government surveillance rights.

We found that the type of surveillance affects its acceptance. Online surveillance encounters more resistance than surveillance in public areas, confirming that citizens are more wary of government scrutiny when it violates the private

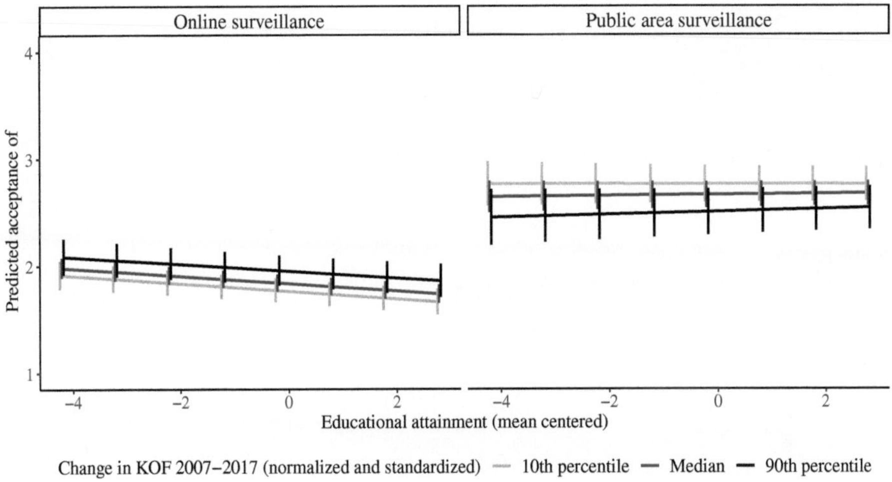

FIGURE 5.5 Predicted Acceptance of Surveillance by education and change in KOF Globalization Index (based on M6 in Table 5.2). Source: EVS (2020). Own calculations.

sphere, aligning with previous findings (cf. Degli Esposti & Santiago Gómez, 2015; van den Broek et al., 2017; Wester & Giesecke, 2019). However, not only do people differentiate among types of surveillance, but also the explanatory mechanisms differ. Most notably, the negative educational gradient in AoS was only found in online settings, with lower educated individuals more willing to allow government surveillance on the communications exchanged on the internet compared to higher educated individuals. Yet, no sizeable educational gradient was found for acceptance of video surveillance in public areas. Hence, despite the strong correlation between the two types of acceptance, it is advised to examine them separately.

Higher educated individuals appear more wary of potentially invasive governmental online surveillance due to their greater awareness of the workings of

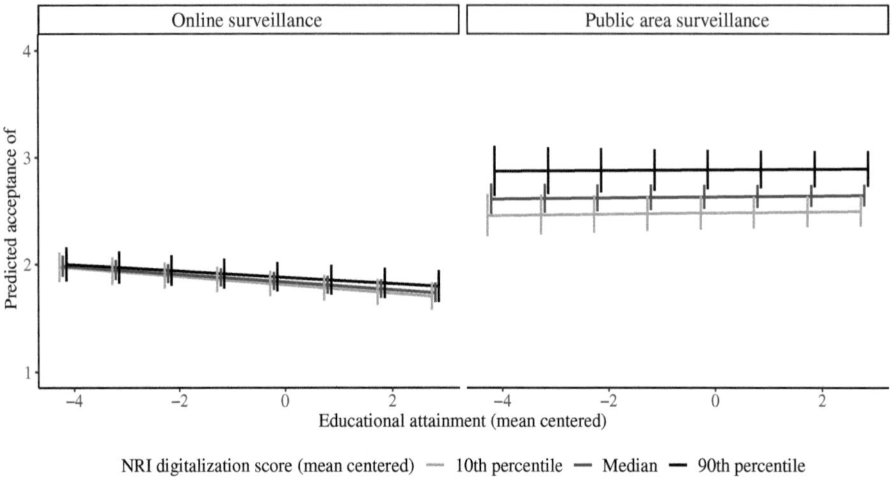

FIGURE 5.6 Predicted Acceptance of Surveillance by education and NRI (based on M7 in Table 5.2). Source: EVS (2020). Own calculations.

modern institutions, confirming the relevance of the reflexive modernization perspective to explain risk stratification at the individual-level, as outlined also in Chapter 1. Against the expectations flowing from reflexive modernization, however, the larger availability of information does not deepen the cleavage between the higher and lower educated strata in society. Nevertheless, the higher AoS in public areas in more digitalized countries, combined with a stable level of AoS online, suggests a sharper differentiation among the types of surveillance in more digitalized countries, which may be explored in future studies.

Results regarding the authoritarian reflex are puzzling. Overall, the preference for a strong leader is positively associated with both types of surveillance, showing the power of the institutional 'safety' narrative. However, this fails to fully explain the educational gradient in AoS, which is mediated when in the online sphere but not in public areas. The rapid expansion of globalization which supposedly drove a security demand among the lower educated individuals is mostly unrelated to AoS. Taken together, these findings lead to discard the expectations derived from the cultural backlash theory. There are, however, some limitations. Regarding measurement, the change in the KOF-GI might not be able to adequately capture the social change that Norris and Inglehart (2019) identified as the spark of the backlash. Future studies may investigate different signals of social change. Second, the cultural backlash might not apply to AoS because of the general securitarian political climate associated with surveillance nowadays. The political landscape of the whole continent, also in countries involved for longer time in globalization processes, has seen an increment in right-wing authoritarian movements who leverage societal threats, e.g., terror attacks occurred in Europe, to promote a securitarian narrative which justifies governmental surveillance efforts, with little room for opposition.

Another methodological limitation in our study is item non-response. Though this problem is not uncommon in studies relying on survey data, it appears particularly problematic when dealing with attitudes towards surveillance, and warrants caution in future studies using these variables.

Generally, as concerns differences among individuals, the educational gradient in AoS online was found to be explained by a combination of factors, with a higher tolerance for online surveillance among lower educated individuals rooted in both a lack of awareness over the functioning of modern institutions and a stronger demand of authoritarianism. Additionally, institutional knowledge was found to be negatively associated with AoS in public areas, indicating that individuals more critical towards modern institutions also perceive manufactured risks stemming from video surveillance systems – which are highly relevant, thinking for instance of systems such as facial recognition technologies. These findings seem to challenge one of the assumptions of the security-privacy trade-off model (van Heek et al., 2017), according to which people are willing to renounce to privacy in exchange for more security. Our study provides some evidence that privacy risks, rather than being willingly accepted, are not fully acknowledged among vulnerable strata of the population. Questioning the accuracy of the assessment of privacy-related risks is an important task for future studies (cf. Marwick & Hargittai, 2018), as it directly challenges the legitimacy of policies regulating the use of new invasive surveillance systems. After all, vulnerable populations also tend to live in areas with higher crime rates, more likely to be surveyed, and feeding a recursive loop, with technology-generated risks made more likely among groups that need protection from safety risks.

Additionally, albeit our results showed differences across countries in the educational gradients in AoS, the two explanations hereby considered - the knowledge gaps enabled by uneven levels of digitalization and the cultural backlash sparked by the expansion of globalization processes - failed to explain them. Future studies should investigate the role of different national characteristics, such as the legacy of authoritarian surveillance (cf. Samatas, 2005) which may have left long-lasting effects on the way citizens assess the government's rights to surveillance.

Our study showed a negative educational gradient in accepting government surveillance of online communications, due to the lack of a reflexive mindset, and – to a lesser extent - stronger securitarian demand among lower educated individuals. The higher tolerance for online government surveillance among lower educated individuals and the risks it entails such as social sorting (Lyon, 2005; Mann & Matzner, 2019), becomes alarming when coupled with previous findings showing how lower educated individuals are less prone to manage their privacy online, as emerged in Chapter 4 and in the study by Y. J. Park and Chung (2017). Future studies should extend the investigations to other dimensions of social vulnerability such as ethnicity, age and social class, so as to remain vigilant of potential vicious circles, whereby vulnerable strata support invasive surveillance policies yet also lack the resources to shield themselves from potential harms, reinforcing inequalities. Our findings underscore the importance of raising awareness about the potential benefits and dangers of online surveillance.

6 Conclusions

A risk society is a society where we increasingly live on a high technological frontier which absolutely no one completely understands and which generates a diversity of possible futures (Anthony Giddens, 1999, p.3)

I • THE DATAFIED RISK SOCIETY

Datafication processes are now embedded in virtually every aspect of life. As an example, among 5.2 billion internet users worldwide, every minute an estimated 5.7 million Google searches occur, and $283,000 are spent on Amazon.[1] Each transaction occurring online, as well as interactions with material objects connected to the internet (e.g. surveillance cameras, smart home devices), generate data flows which are hardly avoidable by individuals. As elaborated in Chapter 1, the constant generation of data about individuals and societal processes enabled by datafication is an opportunity to advance knowledge and efficiency, but it also brings about risks which are difficult to quantify - i.e. privacy risks, digital freedom risks (Beck, 2013). As such, a seemingly innocent action like 'accepting cookies' can have unforeseen repercussions in terms of who gets access to that generated data and why.

In this final chapter, I draw broader conclusions from the analyses presented

[1] See Domo's Data never sleeps infographic, version 9.0 https://www.domo.com/learn/infographic/data-never-sleeps-9 (last accessed on 15-12-2021).

in the empirical chapters. As explained in Chapter 1, the thesis aimed at incorporating datafication processes into Beck's risk society perspective, with a twofold contribution: on the one hand, the risk society concepts can shed better light on datafication; on the other hand, the combination of the two elements offered the opportunity to empirically test some of the claims of the risk society theory. After summarizing the main findings and their theoretical implications, I pinpoint some general limitations (in addition to the ones mentioned in each specific chapter) and, finally, close with some final remarks.

The social acknowledgement of risk

As explained by Beck (1992), many institutions, and the interactions among them, are involved in the reflexive process of production and definition of risks (the 'relations of definition', see Chapter 1). The growing awareness over the limitations of modernity enabled by reflexivity is expected to negatively impact trust in modern institutions which is why, in an attempt to avoid delegitimization, institutions try to deflect their responsibility over the (re)production of risks via organized irresponsibility. This mechanism, however, lacked empirical grounding, which led to the formulation of two questions: first, *to what extent does institutional trust drop when the inherent limits of datafication processes become visible?*; second: *to what extent are individuals aware of the manufactured privacy risks generated by datafication processes?* The two questions are linked to each other as they both shed light on the role of organized irresponsibility from the perspective of its impact on individuals.

With regards to institutional trust, the results presented in Chapter 2 showed how trust in social media – one of the most prominent actors of datafication – does not decrease when individuals are confronted with the notion that their data is easily misused and that they risk to be targeted for anti-democratic purposes, challenging the 'worked and won' dynamic of trust in the risk society. One alternative explanation is the exogenous account of institutional trust (cf. Mishler & Rose, 2001), according to which institutional trust reflects an individual's tendency to trust or their cultural predispositions rather than a critical evaluation of the performance of an institution, and is therefore not adjusted as a consequence of short-term shocks.

Another explanation, however, directly targets the concept of trust in data institutions. As anticipated in Chapter 1, for most 'traditional' abstract systems (e.g. the health care, the government), the trustee can rely on some experience with representatives of these systems (e.g., the GP, the clerk at the administrative office), experience which can be used to infer the quality of the larger system (Giddens, 1990). This is not the case for the big players of datafication, e.g. platforms like social media and other big tech company. There, the only direct interaction a citizen may have is with a virtual interface, which is often populated by content created by other users yet actively re-organized by algorithms designed by the platform themselves, raising questions about the actual object of trust: is it the other users creating and sharing content, is it the underlying technology, or is it the platform as an actual institution? This is a puzzle to be solved in the future.

If trusting a data platform means trusting the other users, then it may resemble a form of dispositional social trust rather than institutional trust. If trusting platforms is a function of their constantly refined technological affordances, then trust is going to reflect the evaluation of the efficiency of the systems: this may explain why, as seen in Chapter 2, trust in social media has not massively dropped as a consequence of Cambridge Analytica. After all, the functionalities of Facebook and other platforms were not affected.

The 'worked and won' dynamic of institutional trust in the risk society emerges once trusting platforms is rooted in an evaluation of their active role in organizing the online discourse. Indeed, platforms thrive by monetizing users' data and producing data-related risks but they are also the same institutions that, in an authentic reflexive dynamic, provide users with access to information and knowledge (Gillespie, 2014). The way information is displayed, ranked, and recommended, is not neutral but depends on the architectural choices and algorithms implemented by the platforms in the background (Beer, 2017). In other words, platforms can act as gatekeepers of the knowledge about the risks they themselves produce: this has to be widely acknowledged before the actual 'benevolence' of platforms can be assessed (cf. Mayer et al., 1995).

Yet, trying to cover this active role in reassembling knowledge can be seen as a part of organized irresponsibility, or the strategies modern institutions adopt in an attempt to mask their role in the production of manufactured risks. Results presented in Chapter 3 – aligning with results from other studies - suggest that when confronted with 'visible' risks such as a pandemic, individuals tend to underestimate the latent risks of using a dataveillance technology, since the privacy-invasive features of a hypothetical COVID-19 Health Pass did not impact its public acceptability in the Netherlands. Coupled with the findings that trust in social media did not drop when privacy risks were unveiled (cf. Chapter 2) and with the relatively high acceptance of government surveillance in public areas despite the risks it entails (cf. Chapter 5), these results suggest the success of the organized irresponsibility strategy, which constitutes an important contribution that the Risk society perspective brings to the understanding of datafication processes.

However, I identified organized irresponsibility only indirectly, by observing the lack of reactions against invasive surveillance activities from an individual perspective. Future research should investigate the ways institutions actively adopt organized irresponsibility as a strategy from an organizational and even technological perspective. While I outlined above the problems related to platforms, it is important to also monitor the developments of 'traditional' institutions. As governmental processes increasingly rely on digital systems removing face-to-face intermediation, and surveillance shifts into dataveillance, organized irresponsibility may interfere with the public institutions' accountability.

The success of organized irresponsibility tactics may also deepen the cleavage between the concerns of experts and those of the populace as regards the risks of datafication, whereas – as explained above – a full acknowledgement of the active role of platforms and institutions in the production of data is needed to really understand their impact. As seen in Chapter 3, the concerns of experts over the privacy intrusiveness of a health surveillance tool seem to be unmatched by concerns among citizens. This will be increasingly problematic as datafication processes expand on a global scale, now even accelerated by the COVID-19 pandemic which justifies extensive health surveillance measures combined with a forced digitalization of many aspects of social life. The risk of surveillance creep (Calvo et al., 2020) warrants active monitoring of these processes from researchers, activists and policy makers.

Organized irresponsibility in the process of social definition of datafication risks also occurs at the level of epistemology, as promoters of datafication processes often rely on dataism, or the belief in the intrinsic value of data (van Dijck, 2014). For instance, Cukier and Mayer-Schoenberger advocate for an epistemological twist in order to fully seize the opportunities of the datafication process:

> Using great volumes of information [...] requires three profound changes in how we approach data. The first is to collect and use a lot of data rather than settle for small amounts or samples, as statisticians have done for well over a century. The second is to shed our preference for highly curated and pristine data and instead accept messiness: in an increasing number of situations, a bit of inaccuracy can be tolerated, because the benefits of using vastly more data of variable quality outweigh the costs of using smaller amounts of very exact data. Third, in many instances, we will need to give up our quest to discover the cause of things, in return for accepting correlations. (Cukier & Mayer-Schoenberger, 2013, p. 29)

Social sciences and theories can help illuminate the shortcomings of this approach, and enable alternative models that allow a more timely acknowledgement of the risks coming from datafication. For instance, more data does not imply better data (Boyd & Crawford, 2012): since social imbalances are built into data (Joyce et al., 2021), having more data does not shield from bias. Relatedly, the complexity embedded in the production of data, starting from the choice of what to (not) measure and arriving to 'read' the patterns found in the data, requires interpretation (Boyd & Crawford, 2012; Kitchin, 2014). This also highlights the weakness of this renewed empiricist approach (Kitchin, 2014) which renounces to explain (the why) and settles with description alone (the what), possibly leading to a flawed and partial understanding of the world. Social scientists should keep contributing to the debate and stress the necessity to recognize that data is not given naturally but is a social product (Joyce et al., 2021) obtained via abstraction processes (Kitchin, 2014; Mejias & Couldry, 2019) and that, as such, the production of data entails risks.

Risk stratification

As seen in Chapter 1, the negative consequences enabled by datafication processes are unevenly distributed in societies (Brayne, 2017; Eubanks, 2018; S. Park & Humphry, 2019). Relatedly, Beck suggested the pivotal role of knowledge to recognize manufactured risks by enabling causal interpretations. Given the mixed empirical findings yielded by previous research, I applied Beck's idea to datafication-induced risked and asked *whether an individual's educational achievement affects their ability to acknowledge datafication-induced risks through a reflexive mindset, or lack thereof.*

My investigation focused mostly on the role of formal education as a form of knowledge. In Chapters 4 and 5, I showed how higher educated individuals are more aware of privacy risks involved in data exchanges, which is reflected in their higher tendency of protecting e-privacy and in their lower tolerance of online surveillance. Additionally, in Chapter 2, higher educated individuals were found to be more wary of social media. These findings underscore the pivotal role of higher educational institutions in shaping digital-savvy citizens who are better equipped to face the challenges of datafication, which is a further dimension of comparative advantage of the higher educated over the lower educated in information-intense societies.

The results, all in all, confirm that knowledge matters as an element of risk stratification, given that not only education but also digital skills (as a form of informal knowledge) and institutional knowledge explain a more wary attitude towards datafication. However, only part of this knowledge stratification is based on a heightened awareness over the limitations of modernity, as when the mechanisms related to reflexive mindset are tested, the situation becomes nuanced. In the case of e-privacy management, reflexive mindset explains a small portion of the educational gap (cf. Chapter 4). As for acceptance of online surveillance, institutional knowledge (as a proxy of reflexive mindset) does indeed explain the negative educational gradient, but so do – though to a lesser extent – also authoritarian attitudes (cf. Chapter 5).

Going beyond the limitations reported in section 2, this finding reinforces the strength of the organized irresponsibility dynamic, given that even higher educated individuals seem unlikely to attribute datafication-induced risks to the responsible institutions. This may also be due to a popular narrative surrounding privacy risks that sees them rooted in individual responsibility, the 'nothing to hide' argument (cf. Solove, 2007): people should not worry about data breaches as long as they have not produced any compromising evidence. This is a severe underestimation of the power of datafication, which enables the reconstruction of sensitive information based on seemingly uncompromising data (e.g. inferring race based on the writing style, see the Gizmodo example in Chapter 1), therefore placing the control of the data outside the hands of the data subject.

The emergence of a reflexive mindset at the individual level is also affected by contextual factors such as, I argued, the availability of ICTs, which led me to ask *to what extent do educational gaps in the acknowledgement of datafication-induced risk vary by the levels of digitalization in different European countries?* The analyses presented in Chapter 4 indicate that the educational divide in e-privacy narrows in more digitalized countries, suggesting that the educational gaps in the acknowledgement of datafication-induced risk do vary. However, the variation is not in the direction prospected by the risk society theory, as the educational gap narrows rather than widening. Additionally, in Chapter 5 I showed how the educational gradients in acceptance of surveillance do not vary by the level of digitalization of countries. In a nutshell, the level of a digitalization in a country does not exacerbate the knowledge-based risk stratification.

The lack of support of the reflexive modernization thesis at the level of differences across countries may be linked to the use of digitalization. On the one hand, the tendency to converge towards high levels of digitalization – also supported by funding and action plans of the European Union – may blur differences across countries. On the other hand, it may be that the spread of ICTs does not really enable the uncovering of the risks ICTs institutions themselves produce. Phenomena like echo chambers, i.e. online users being exposed to content which reinforces their pre-existing beliefs and opinions, and recommender systems, i.e. the way big tech platforms decide how to display content, shape the opportunities of new knowledge to emerge and be picked up by individuals. Though this is not something extraneous to Beck's Risk society (e.g. the role of media is considered in the 'relations of definitions'), the extent to which this is changing with digital media is not fully explored yet, as also suggested in the previous section.

Despite the global nature of datafication-induced risks, as also highlighted by Beck (1992, 2002) some regions of the world are worse off than others, and indeed differences across countries in the awareness over datafication risks emerged also in my studies (cf. Chapters 4 and 5). However, given that digitalization failed to explain these differences, other structural factors should be considered. For instance, future studies may investigate to what extent cultural factors, such as trust and post-materialism, foster - or hinder - the emergence of knowledge over datafication risks.

Whereas in accordance with Beck's perspective education matters for the stratification of risk at the individual level, the potential of knowledge to really break the risk production cycle at the societal level by identifying its causes is not yet realized. I therefore propose at least two main directions to follow to improve the understanding of inequalities in the datafied society. First, it should be investigated how 'traditional' sources of inequalities – e.g., age, sex, social class, ethnicity, area of residence, occupational status – add to and complement educational gaps in understanding inequalities in the datafied society. Such an endeavor allows to contribute to the broader debate on the role of ascribed social position vs. the individualization of experience[2] in shaping inequalities in Beck's perspective (cf. Curran, 2013), but also to better understand the consequences of datafication for specific social groups and throughout the life-course (cf., for instance, Mascheroni, 2020, on datafication and childhood).

Second, the peculiar characteristic of datafication and the specialistic skillset (e.g. computational skills) needed to understand its inner workings and causal implications, may generate new social cleavages. Some talk about a 'big data divide' (Andrejevic, 2014; McCarthy, 2016) that has to do with the asymmetries between three classes ('data-classes') of people and organizations (Manovich, 2011): those who generate data, intentionally and not, with their online activities; those who can afford to collect the data; and finally those who have the technical knowledge to analyze them. Future research should monitor

[2] Data-driven systems and algorithms contribute to fostering specialization and individualization by tailoring users' experience to their previous behaviors (e.g. filter bubbles, cf. Gilbert, 2018), creating an interesting contact point to be further explored in the future.

the definition of and interactions between these 'data-classes', and how they relate to other sources of inequalities amidst datafication processes in a global perspective (cf. Burrell & Fourcade, 2020, on the emergence of the 'coding elite' as the new powerful occupational class of surveillance capitalism).

2 • LIMITATIONS AND FUTURE WORK

Beyond the limitations addressed in each chapter, some general limitations on the comparability of results across chapters and elements that are left out from the thesis should be pointed out. In parallel, suggestions for new avenues of research are suggested.

A recurrent limitation across the chapters is the unclear measurement of reflexivity at the individual level, which partially hinders the comparability of some of the results on mechanisms behind the educational gaps. For instance, while in Chapter 4 I used a measure of 'feeling responsible for climate change' to indicate the awareness over the man-made nature of modern risks, in Chapter 5 I used 'Institutional knowledge'. While this is partially an unescapable consequence of using secondary data, it is also partially rooted in the lack of clarity on how to operationalize the concept of reflexive mindset, manifest also in previous empirical studies (see overview in Chapter 1). Future work should concentrate on validating these measures or devising new ones that adequately capture the concept.

Relatedly, whereas the availability of survey data including questions related to data, privacy and surveillance is increasing, these often occur in topical surveys (e.g. Eurobarometer) which therefore limits the array of variables included and, consequently, the possibility to test mechanisms at the individual level. Survey data also entails potential limitations in terms of representativeness of the samples and therefore the impossibility to adequately investigate some groups in society (i.e. migrants, often excluded due to language barriers).

There are many ways to overcome these obstacle in future research. First, due to the centrality of datafication processes in contemporary societies, it is recommended that general social surveys devote space to these topics in the future, allowing to link privacy- and surveillance- related questions to a more diverse range of attitudes and behaviors. Second, novel, privacy-preserving, methods can be adopted. For instance, data donations (cf. Araujo et al., 2022; Boeschoten et al., 2022) enable a in-depth analysis of digital behaviors through data download packages (DDPs) while potentially empowering the individuals who, through the donation of their DDP, acquire awareness over the data that is generated about themselves. Third, a qualitative approach could be fruitfully employed to investigate the more in-depth mechanisms underlying human action. For instance, in Chapter 2 and 4 I briefly touched upon the issue of privacy paradox, namely a gap between awareness over privacy risks and actual protection against them (Acquisti et al., 2015; Kokolakis, 2017). According to this framework, awareness of datafication-induced risk might not be sufficient to shield from them, thus potentially limiting the benefits of awareness campaigns and knowledge building. More research is needed to solve the puzzle, and qualitative insights might help to appreciate the contextual nature of privacy protection decisions in the face of datafication.

Another general limitation of the thesis is the limited geo-spatial scope. It should be noted that the European legislative framework, compared to other parts of the world, is more protective of citizens' rights in the face of datafication processes as seen, for instance, with the introduction of the EU General Data Protection Regulation (EU-GDPR). This limits the generalizability of the results to different contexts, as citizens' attitudes towards privacy and surveillance, as well as their awareness over privacy risks, may be affected by the regulatory framework. More generally, the focus on Europe is partially justified by the fact that the shift from the simple to the reflexive modernity envisaged by Beck is dependent upon the existence of the first, simple modernity itself. However, this leads to perpetuate the admittedly Western-centric perspective of Beck's theory (Beck et al., 2003, p. 7) and to underappreciate the global nature of datafication processes, which should be further explored in future studies. We propose here two directions: first, when it comes to studying the role of institutions and organised irresponsibility, the transnational nature of data flows should be considered. Second, contextual specificities in terms of culture, politics, and history, may have a strong impact on the way individuals are aware of, and react to, datafication, and looking beyond Europe would certainly offer more insights.

Among the themes that are not addressed in this thesis, a relevant one is that of voluntary dataveillance. Here, I have concentrated on the consequences of data processes which are somehow imposed top-down by institutions involved in datafication. This is not an extensive account, given that in the era of personal digital data, individuals often engage voluntarily with surveillance practices, e.g. by monitoring themselves using sensors (e.g. Health apps), or monitoring others on social media (cf., e.g., Lupton & Michael, 2017). The perceptions of risks deriving from these processes against the individual gains constitutes an important dimension to understand the awareness over the manufactured risks of datafication.

3 • CONCLUDING REMARKS

Datafication facilitates operations which are fundamental for the functioning of society – e.g. interpersonal communications, access to services, counter-terrorism, etc. As such, datafication processes are set to expand, over time and across geographical areas, challenging the definition of what can be done with data, how, and by whom. At the same time, societies will be increasingly faced with the risks datafication induces. If citizens keep 'accepting cookies' uncritically and underestimate the manufactured risks stemming from datafied practices, it will be more and more difficult to hold responsible institutions accountable, as they will attempt to enact organized irresponsibility to avoid delegitimization. This thesis has shown the key role of education and knowledge to break this circle and be more conscious of privacy risks, showing the potential of increasing awareness around these topics to empower individuals in the face of the datafied risk society.

Appendix to In Zuck We Trust?

The Sources of Trust in Social Media in Times of Data Privacy Controversies

TABLE A.1 Descriptives of all trust-items.

Trust-item	N	Mean	Std. Deviation
Church	1,475	1.91	0.82
Armed forces	1,457	2.60	0.66
Education system	1,484	2.82	0.56
The press	1,493	2.25	0.69
Trade unions	1,442	2.34	0.67
The police	1,496	2.79	0.62
Parliament	1,480	2.35	0.69
Civil service	1,477	2.33	0.63
Social security system	1,478	2.57	0.67
European Union	1,472	2.20	0.76
United Nations organization	1,450	2.42	0.74
Health care system	1,500	2.87	0.65
Justice system	1,463	2.63	0.71
Major companies	1,458	2.26	0.64
Environmental organizations	1,465	2.45	0.73
Political parties	1,488	2.05	0.62
Government	1,504	2.41	0.66
Social media	1,504	1.91	0.60

Question 'How much confidence in [institution]?' | Scale range: 1–4

TABLE A.2 Descriptives of the continuous variables.

Variable	Range	Mean	Std Deviation
Trust in social media (pre-test)	1-4	1.90	0.60
Trust in social media (post-test; N = 1,097)	1-4	1.94	0.55
Trust in government	1-4	2.41	0.65
Institutional trust	1-4	2.40	0.44
Follow politics on social media	1-5	2.29	1.30
Follow politics on tv	1-5	3.26	1.34
Age	18-82	55	16.9
Income	1-10	5.60	2.58
Levels of education	0-7	4.35	1.99

N = 1,504, except for trust in social media obtained from the post-test data.

TABLE A.3 Descriptives of the categorical variables.

Variable	Range	Frequency	Percentage
Change trust in social media (N = 1,097)	Decrease	241	22.0
	No change	560	51.0
	Increase	296	27.0
Social trust	Cannot be too careful	527	35.0
	Most people can be trusted	977	65.0
Post-materialism	Materialism	251	16.7
	Post-materialism	252	16.8
	Mixed	1.001	66.6
Income non-response dummy	Substantial response	1.294	86.1
	Nonresponse	210	13.9
Gender	Men	709	47.1
	Women	795	52.9
Employment status	Employed	730	48.5
	Unemployed	50	3.3
	Student	57	3.8
	Retired	513	34.1
	Other status	154	10.2
Matrix-design	Group 1 (Blocks AB-CD)	247	16.4
	Group 2 (Blocks AC-BD)	255	16.9
	Group 3 (Blocks AD-BC)	250	16.6
	Group 4 (Blocks BC-AD)	244	16.2
	Group 5 (Blocks BD-AC)	244	16.2
	Group 6 (Blocks CD-AB)	264	17.5

N = 1,504, except for trust in social media obtained from the post-test data.

TABLE A.4 Ordinal logistic regression of trust in social media on relevant covariates, including trust in government instead of institutional trust.

Independent variable	Model 2: Trust-nexus	Model 4: Full model
Follow politics on social media[a]		0.334***
		(0.045)
Trust in government[a]	0.598***	0.682***
	(0.087)	(0.089)
Social trust[a]	0.027	0.127
	(0.119)	(0.121)
Postmaterialism (Ref: Mixed)		
- Materialism		0.431**
		(0.151)
- Post-materialism		0.035
		(0.150)
Levels of education[a]		−0.158***
		(0.032)
Age[a]	−0.001	0.000
	(0.005)	(0.006)
Woman (Ref = Man)	−0.016	−0.016
	(0.111)	(0.112)
Income textsuperscripta	−0.023	0.016
	(0.022)	(0.023)
Income missing dummy	0.127	−0.010
	(0.156)	(0.159)

Table continues on next page.

Work status (Ref: Employed)		
- Unemployed	0.901^{**}	0.800^{*}
	(0.310)	(0.314)
- Student	−0.175	−0.186
	(0.316)	(0.321)
- Retired	0.525^{**}	0.510^{**}
	(0.184)	(0.186)
- Other work status	0.504^{*}	0.335
	(0.196)	(0.200)
Follow politics on television[a]	-0.092^{*}	-0.152^{**}
	(0.044)	(0.048)
Cut-off trust in social media = 1	-1.078^{***}	-1.110^{***}
	(0.165)	(0.171)
Cut-off trust in social media = 2	2.200^{***}	2.342^{***}
	(0.175)	(0.182)
Cut-off trust in social media = 3	5.577^{***}	5.766^{***}
	(0.389)	(0.393)
Observations	1.504	1.504

Entries represent ordered log-odds regression coefficients (standard errors in parentheses). Analysis controlled for matrix design. Source: EVS 2017 Netherlands.
[a] Variable centered around the mean | $^{*}p < 0.05$; $^{**}p < 0.01$; $^{***}p < 0.001$

TABLE A.5 Multinomial logistic regression of change in trust in social media (Reference = No Change), using trust in government instead of institutional trust.

Independent variable	Decreased trust	Increased trust
Follow politics on social media[a]	0.126*	−0.087
	(0.064)	(0.062)
Trust in government[a]	0.253	−0.161
	(0.130)	(0.121)
Social trust[a]	0.083	−0.131
	(0.180)	(0.167)
Materalism index (Ref: Mixed)		
- Materialist	0.477*	0.393
	(0.213)	(0.206)
- Postmaterialist	0.172	0.154
	(0.220)	(0.207)
Educational level[a]	−0.023	0.069
	(0.047)	(0.044)
Age[a]	0.002	0.001
	(0.008)	(0.008)
Female	−0.270	−0.038
	(0.166)	(0.155)
Income[a]	−0.017	−0.022
	(0.035)	(0.033)
Income missing	0.157	0.513*
	(0.234)	(0.208)

Table continues on next page

Work status (Ref: Employed)		
- Unemployed	−0.064	−1.151*
	(0.415)	(0.520)
- Student	−1.123	−0.479
	(0.595)	(0.443)
- Retired	−0.070	−0.557*
	(0.271)	(0.251)
- Other work status	−0.341	−0.708*
	(0.296)	(0.285)
Follow politics on television[a]	−0.049	0.212**
	(0.071)	(0.066)
Intercept	−1.766**	−0.638
	(0.636)	(0.584)
Nagelkerke's R2	0.069	
Observations	1,097	

Entries represent ordered log-odds regression coefficients (standard errors in parentheses). Analysis controlled for matrix design. Source: EVS 2017 Netherlands.
[a] Variable centered around the mean | $^{*}p < 0.05$; $^{**}p < 0.01$; $^{***}p < 0.001$

Appendix to Public acceptance of a COVID-19 Health Pass

Evidence from a vignette study in the Netherlands

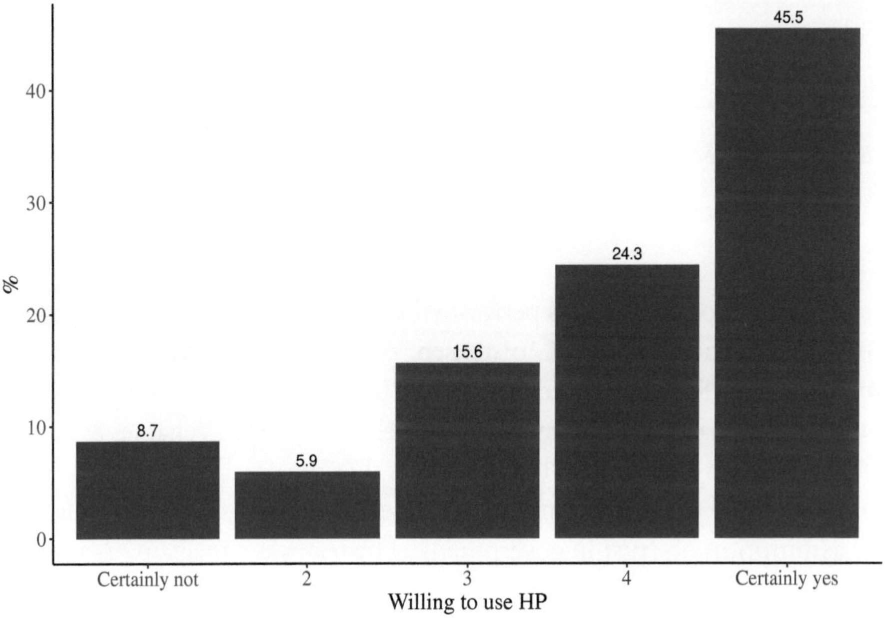

FIGURE B.1 Frequency distribution of willingness to use the COVID-19 HP, N = 1,454.

I • TEST OF PROPORTIONAL ODDS

TABLE B.1 Testing the proportional odds assumption.

	LR test					Brant test		
	Df	logLik	AIC	LRT	p-value	χ^2	df	p-value
omnibus		-1863.6	3749.2			58.7	21	
Trust in government	3	-1854.2	3736.4	18.7845	0.000***	13.2	3	0.000***
Trust in science	3	-1859.8	3747.6	7.5508	0.056	6.3	3	0.1
Concern over coronavirus	3	-1861.7	3751.3	3.8514	0.277	4.1	3	0.25
Vaccination hesitant	3	-1858.1	3744.2	10.9869	0.012*	13.9	3	0.000***
Female	3	-1862.1	3752.3	2.9218	0.404	3.6	3	0.31
Age	3	-1858.3	3744.5	10.6753	0.013*	7.5	3	0.06
Educational attainment	3	-1859.8	3747.7	7.5155	0.057	2.05	3	0.56

Df = degrees of freedom; logLik = log Likelihood; AIC = Akaike Information Criterion; LRT = Likelihood Ration Test; p = p-value; χ^2 = Chi-square;
$^{*} p < 0.05$; $^{**} p < 0.01$; $^{***} p < 0.001$

The likelihood ratio test is performed with the function `nominal test` from the `ordinal` package (Christensen, 2019), and the brant test (from the `brant` package Schlegel & Steenbergen, 2020) is performed on the same model, yet estimated via the `polr` function in the `MASS` package (Venables & Ripley, 2002). For both tests, the null hypothesis is that the parallel lines assumption holds. The two tests yield similar results (see Table B.1) and lead to relax the PO assumption on trust in government. The test was run again on a model on which the PO assumption on trust in government was relaxed, and also on models including the variables related to the experimental conditions, and for the other main independent variable included in the study, i.e. trust in science, the PO assumption looks tenable.

2 • QUESTION WORDING

The wording of the 24 vignettes is reported in this section. The original wording is as follows (with experimental conditions between square brackets):

> Om de coronacrisis te bestrijden, wordt er gewerkt aan een testbewijs. In onderstaande tekst beschrijven we hoe een testbewijs eruit zou kunnen zien. Kunt u aangeven of u dit testbewijs zou gebruiken, en of u denkt dat het werkt om de pandemie onder controle te krijgen?
>
> Het coronatestbewijs waaraan gewerkt wordt heeft als doel om [A1 te voorkomen dat mogelijk geïnfecteerde mensen toegang hebben tot drukbezochte plekken | A2 publieke plekken veilig te maken voor iedereen]. De toegang tot deze plekken is alleen mogelijk na controle van een geldige negatieve coronatest of met vaccinatiedata die worden bijgehouden [B1 door de overheid | B2 door uw zorgverzekeraar]. Deze gegevens worden anoniem aangeboden in de vorm van een [C1 code op papier | C3 digitale code via een app of SMS]. Om fraude te voorkomen [D1+D2 worden | D3 wordt] ter controle via uw identiteitskaart of paspoort ook [D1 uw initialen en geboortedatum | D2 uw volledige naam en geboortedatum | D3 uw burgerservicenummer] vermeld op het testbewijs.

An example of a vignette in the original language (Dutch) is displayed in Figure B.2.

The translation is as follows (with experimental conditions between square brackets):

> Finally. In order to combat the coronacrisis, a certificate is being

worked on. In the text below we describe how a certificate could look like. Could you indicate afterwards whether you would use this certificate, and if you think it would be effective to get the pandemic under control?

The purpose of the corona certificate that is being worked on is [**A1** to prevent potentially infected people from having access to busy places | **A2** to make public places safe for everybody]. Access to these places is only possible after checking a valid negative corona test or with vaccination data that is kept [**B1** by the government | **B2** by your insurance provider]. This data is provided anonymously in the form of a [**C1** code printed on paper | **C3** digital code via an app or text]. To prevent fraud, [**D1** your initials and date of birth | **D2** your full name and date of birth | **D3** your citizen service number] will also be stated on the certificate for verification via your identity card or passport.

Tot slot.

Om de coronacrisis te bestrijden, wordt er gewerkt aan een testbewijs. In onderstaande tekst beschrijven we hoe een testbewijs eruit zou kunnen zien. Kunt u aangeven of u dit testbewijs zou gebruiken, en of u denkt dat het werkt om de pandemie onder controle te krijgen?

Het coronatestbewijs waaraan gewerkt wordt heeft als doel om te voorkomen dat mogelijk geïnfecteerde mensen toegang hebben tot drukbezochte plekken. De toegang tot deze plekken is alleen mogelijk na controle van een geldige negatieve coronatest of met vaccinatiedata die worden bijgehouden door de overheid. Deze gegevens worden anoniem aangeboden in de vorm van een code op papier. Om fraude te voorkomen wordt ter controle via uw identiteitskaart of paspoort ook uw burgerservicenummer vermeld op het testbewijs.

Zou u dit testbewijs gebruiken?

Heel zeker niet ○ ○ ○ ○ ○ Heel zeker wel

Denkt u dat dit testbewijs werkt om de pandemie onder controle te krijgen?

Helpt heel zeker niet ○ ○ ○ ○ ○ Helpt heel zeker wel

Vorige | Verder

LISS PANEL

TILBURG UNIVERSITY

FIGURE B.2 Example of vignette.

TABLE B.2 Question wording and answer categories per variable.

Variable	Question wording (original)	Answer categories (original)	Question wording (English)	Answer categories (English)
Willingness to use HP	Zou u dit testbewijs gebruiken?	1. Heel zeker niet … 5. Heel zeker wel	Would you use such a certificate?	1 Certainly not … 5 Certainly yes
Trust in government (reverse coded)	Hoeveel vertrouwen hebt u in… Overheid	1. Heel veel; 2. Tamelijk veel ; 3. Niet zo veel; 4. Helemaal geen	How much confidence do you have in… the government	1. A great deal; 2. Quite a lot; 3. Not so much; 4. None at all
Trust in science (reverse coded)	Hoeveel vertrouwen hebt u in… De wetenschap	1. Heel veel; 2. Tamelijk veel; 3. Niet zo veel; 4. Helemaal geen	How much confidence do you have in… science	1. A great deal; 2. Quite a lot; 3. Not so much; 4. None at all
Concern over coronavirus	In hoeverre maakt u zich zorgen over de coronacrisis in het algemeen?	1. Helemaal niet; 2. Niet veel; 3. Een beetje; 4. Veel; 5. Heel veel	To what extent are you concerned about the coronacrisis in general?	1. Not at all; 2. Not much; 3. A bit; 4. Much; 5. Very much

Table continues on next page

<table>
<tr><th>Variable</th><th>Question wording (original)</th><th>Answer categories (original)</th><th>Question wording (English)</th><th>Answer categories (English)</th></tr>
<tr><td>Vaccination intention (target variable: vaccination hesitant)</td><td>Zou u zich laten vaccineren tegen COVID-19?</td><td>1. Ik ga me zeker niet laten vaccineren; 2. Ik ga me waarschijnlijk niet laten vaccineren | 3. Ik ga me waarschijnlijk wel laten vaccineren; 4. Ik ga me zeker wel laten vaccineren; 5. Ik ben al gevaccineerd of ik heb een vaccinatieafspraak</td><td>Would you get vaccinated against COVID-19?</td><td>1. I will certainly not get vaccinated; 2.I will probably not get vaccinated | 3. I will probably get vaccinated; 4. I will certainly be vaccinated; 5. I have been vaccinated already or I have an appointment</td></tr>
<tr><td>Female</td><td>Geslacht</td><td>1. Man; 2. Vrow</td><td>Gender</td><td>1. Man; 2. Woman</td></tr>
<tr><td>Educational level</td><td>Hoogste opleiding met diploma</td><td>1. basisonderwijs; 2. vmbo; 3. havo/vwo; 4. mbo; 5. hbo; 6. wo</td><td>Highest educational level with diploma</td><td>1. Primary school; 2. pre-vocational education ; 3. general secondary school ; 4. middle-level applied education; 5. University of applied sciences; 6. University</td></tr>
<tr><td>Age</td><td>Geboortejaar</td><td></td><td>Year of birth</td><td></td></tr>
</table>

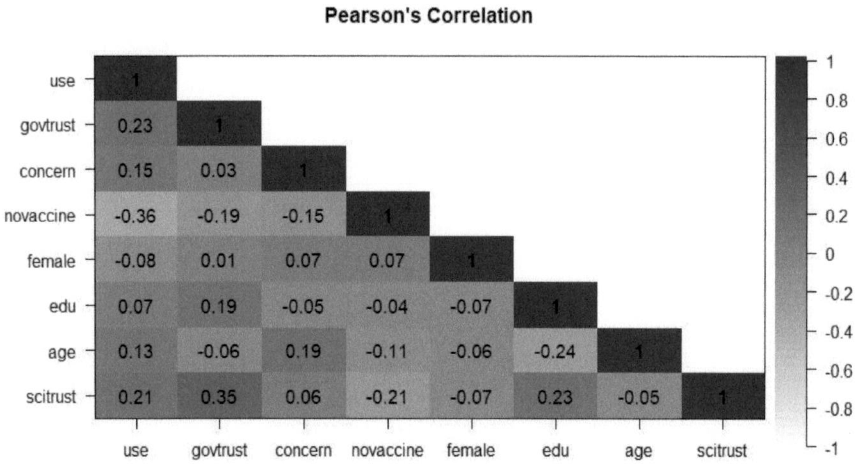

FIGURE B.3 Correlation matrix of variables included in the regression models.

Note: use = Willingness to use HP; govtrust = trust in government; concern = concern over coronavirus; novaccine = vaccine hesitant; edu = education; scitrust = trust in science.

3 • ADDITIONAL RESULTS

TABLE B.3 Partial proportional odds models of Willingness to use the COVID-19 HP by individual characteristics and experimental conditions, with interaction terms.

	Model 1	Model 2	Model 3	Model 4	Model 5	Model 6	Model 7	Model 8	Model 9
Trust in science	0.34***	0.35***	0.34***	0.16	0.36**	0.55***	0.34***	0.34***	0.34***
	(0.09)	(0.09)	(0.09)	(0.12)	(0.12)	(0.14)	(0.09)	(0.09)	(0.09)
Concern over coronavirus	0.22***	0.22***	0.22***	0.22***	0.22***	0.23***	0.22***	0.22***	0.22***
	(0.06)	(0.06)	(0.06)	(0.06)	(0.06)	(0.06)	(0.06)	(0.06)	(0.06)
Vaccination hesitant	−1.66***	−1.66***	−1.67***	−1.64***	−1.66***	−1.66***	−1.65***	−1.66***	−1.66***
	(0.19)	(0.19)	(0.19)	(0.18)	(0.18)	(0.18)	(0.18)	(0.18)	(0.18)
Female	−0.22*	−0.21*	−0.23*	−0.22*	−0.22*	−0.23*	−0.22*	−0.22*	−0.22*
	(0.10)	(0.10)	(0.10)	(0.10)	(0.10)	(0.10)	(0.10)	(0.10)	(0.10)
Age[a]	0.02***	0.02***	0.02***	0.02***	0.02***	0.02***	0.02***	0.02***	0.02***
	(0.00)	(0.00)	(0.00)	(0.00)	(0.00)	(0.00)	(0.00)	(0.00)	(0.00)
Educational attainment[a]	0.04	0.04	0.03	0.04	0.04	0.04	0.04	0.04	0.04
	(0.04)	(0.04)	(0.04)	(0.04)	(0.04)	(0.04)	(0.04)	(0.04)	(0.04)
Experimental conditions									
Purpose: ensure safe access	0.15	0.17	−0.05						
	(0.14)	(0.14)	(0.17)						
Recipient: own insurer	0.09			−0.81*			−0.34		
	(0.14)			(0.36)			(0.21)		
Support: digital		0.10			0.01			0.02	
		(0.14)			(0.36)			(0.21)	
Attribute: full name			−0.15			0.71			−0.06
			(0.17)			(0.46)			(0.27)
Attribute: BSN			−0.13			0.51			0.12
			(0.17)			(0.43)			(0.26)
Interaction terms:		*x Purpose: ensure safe access*			*x trust in science*			*x trust in government*	
Recipient: own insurer	−0.29			0.34*			0.22		
	(0.20)			(0.16)			(0.14)		
Support: digital		−0.32			−0.03			−0.06	
		(0.20)			(0.16)			(0.14)	
Attribute: full name			0.28			−0.33			0.04
			(0.25)			(0.20)			(0.18)
Attribute: BSN			−0.08			−0.31			−0.21
			(0.24)			(0.19)			(0.17)

(Table continues to next page)

	Model 1	Model 2	Model 3	Model 4	Model 5	Model 6	Model 7	Model 8	Model 9
Thresholds: Intercept									
1|2	-0.75^{*}	-0.71^{*}	-0.89^{**}	-1.22^{***}	-0.79^{*}	-0.40	-0.98^{**}	-0.79^{**}	-0.78^{*}
	(0.30)	(0.30)	(0.31)	(0.34)	(0.33)	(0.38)	(0.31)	(0.30)	(0.32)
2|3	-0.16	-0.12	-0.30	-0.62	-0.20	0.20	-0.38	-0.20	-0.18
	(0.29)	(0.29)	(0.30)	(0.33)	(0.32)	(0.37)	(0.29)	(0.29)	(0.31)
3|4	0.80^{**}	0.84^{**}	0.66^{*}	0.34	0.76^{*}	1.16^{**}	0.58^{*}	0.76^{**}	0.77^{*}
	(0.28)	(0.28)	(0.29)	(0.31)	(0.31)	(0.37)	(0.28)	(0.28)	(0.30)
4|5	1.55^{***}	1.59^{***}	1.41^{***}	1.09^{***}	1.51^{***}	1.91^{***}	1.33^{***}	1.51^{***}	1.52^{***}
	(0.28)	(0.28)	(0.29)	(0.31)	(0.31)	(0.37)	(0.28)	(0.28)	(0.30)
Thresholds: Trust in government									
1|2	-0.75^{***}	-0.76^{***}	-0.76^{***}	-0.75^{***}	-0.75^{***}	-0.75^{***}	-0.63^{***}	-0.78^{***}	-0.82^{***}
	(0.15)	(0.15)	(0.15)	(0.15)	(0.15)	(0.15)	(0.17)	(0.16)	(0.18)
2|3	-0.64^{***}	-0.65^{***}	-0.65^{***}	-0.64^{***}	-0.64^{***}	-0.64^{***}	-0.52^{***}	-0.67^{***}	-0.71^{***}
	(0.12)	(0.12)	(0.12)	(0.12)	(0.12)	(0.12)	(0.14)	(0.14)	(0.16)
3|4	-0.52^{***}	-0.53^{***}	-0.52^{***}	-0.52^{***}	-0.52^{***}	-0.52^{***}	-0.41^{***}	-0.55^{***}	-0.59^{***}
	(0.09)	(0.09)	(0.09)	(0.09)	(0.09)	(0.09)	(0.12)	(0.12)	(0.14)
4|5	-0.22^{**}	-0.22^{**}	-0.22^{**}	-0.22^{**}	-0.22^{**}	-0.22^{**}	-0.11	-0.25^{*}	-0.28^{*}
	(0.08)	(0.08)	(0.08)	(0.08)	(0.08)	(0.08)	(0.11)	(0.11)	(0.13)
Model fit									
AIC	3739.93	3739.34	3741.61	3735.36	3739.94	3738.38	3737.76	3739.80	3739.51
BIC	3829.73	3829.14	3841.97	3819.87	3824.46	3833.46	3822.27	3824.31	3834.58
Log Likelihood	−1852.97	−1852.67	−1851.81	−1851.68	−1853.97	−1851.19	−1852.88	−1853.90	−1851.75
Num. obs.	1454	1454	1454	1454	1454	1454	1454	1454	1454

Log odds (standard error in parenthesis); [a] Median-centered
$^{***}p<0.001$; $^{**}p<0.01$; $^{*}p<0.05$;

Appendix to The Closing Educational Gap in E-privacy Management in European Perspective

I • NON-INTERNET USERS

Based on the index of internet use provided in the dataset, there are overall 6,451 respondents who declared they either never access the internet (5,942 respondents) or that they have no access to the internet at all (509 respondents). These respondents were not asked questions on e–privacy, and are thus excluded from the analyses presented in the manuscript.

Figure C.1 displays the prevalence of non–internet users in each country. In Lithuania, Romania and Malta, almost two respondents out of five do not access the internet, whereas in the Netherlands, Sweden and Denmark, it is less than one respondent out of ten.

Table C.1 presents the results of a multilevel linear probability model (details on the operationalization of the predictors can be found in the manuscript). The Intraclass Correlation Coefficient of the empty model is 0.05, meaning that 5% of the variation in non–internet use can be attributed to variation between countries. In line with the digital divide literature, we find that the lower educated, women, elderly, unemployed and people living in rural areas have a higher likelihood of not using the internet. Surprisingly, also being a student is associated with a higher likelihood of not using the internet. There

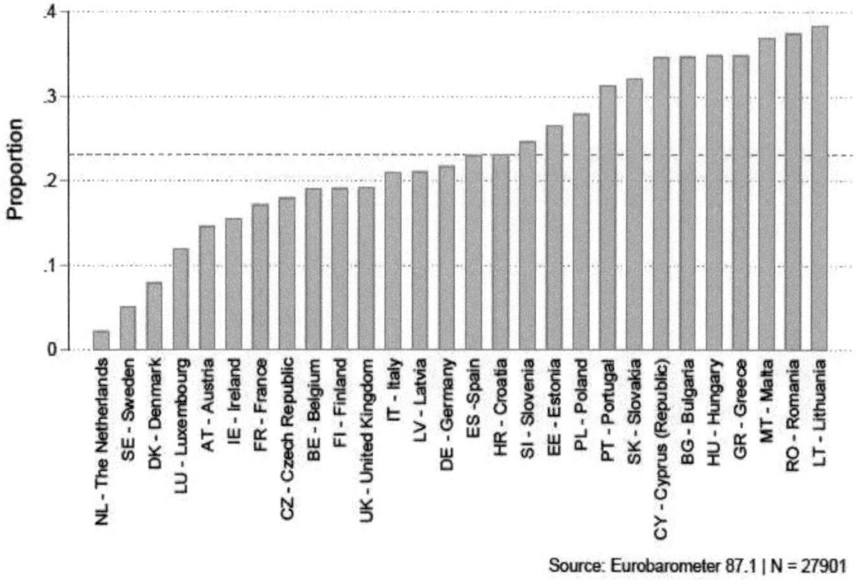

FIGURE C.1 Proportion of non–internet users by country. The dotted line represents the grand mean across countries.

is some variation in the impact that years spent in education has on internet use across countries. In particular, the cross-level interaction between the level of digitalization and years spent in education presented in the last model shows that the negative impact of years spent in education on the chance of not using the internet is smoothened by a more advanced stage of digitalization. In other words, the likelihood of lower educated people of not using the internet, compared to higher educated, is stronger in countries with a low degree of digitalization; on the contrary, in countries with a high degree of digitalization, the impact of years spent in education on frequency of internet use is almost non-existent (see Figure C.2).

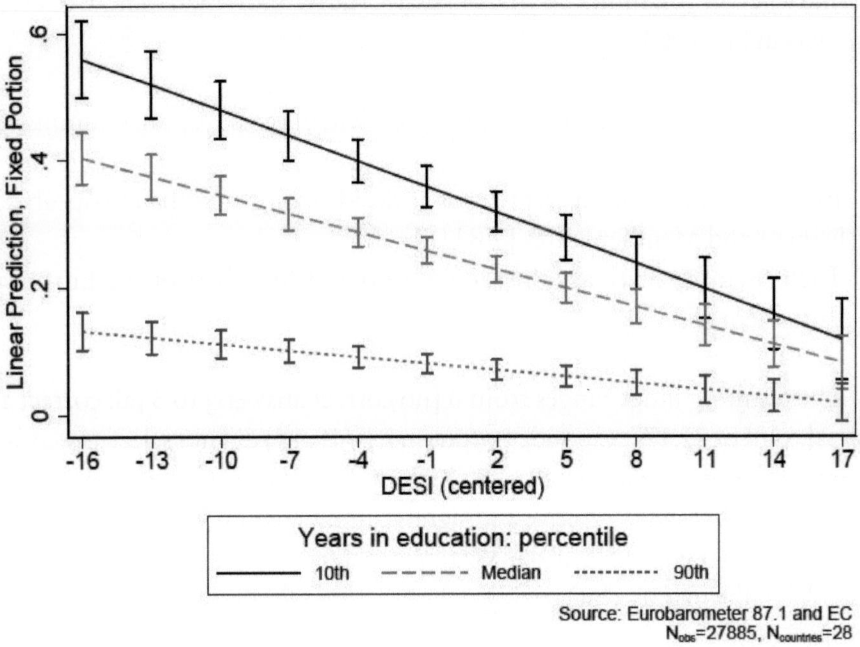

FIGURE C.2 Predicted values of non-internet use and 95% confidence intervals by 10th, 50th and 90th percentile of years spent in full-time education and DESI.

2 • ALTERNATIVE OPERATIONALIZATION OF REFLEXIVE MINDSET

As a robustness check, analyses at the individual levels were repeated using an alternative operationalizazion of reflexive mindset. Previous research indicated that institutional knowledge is also an aspect of a reflexivity at the individual level (Achterberg et al., 2017). We hence computed an index of Institutional knowledge by counting the number of correct answers given to the following question: For each of the following statements about the EU could you please tell me whether you think it is true or false? True – False – Don't know [Correct answers in brackets]

A The European Parliament elects the President of the European Commission [TRUE]
B In the European Union, legislation is decided jointly by the European parliament and Member States [TRUE]
C Each Member State has the same number of Members of the European Parliament [FALSE]

The resulting index ranges from 0 [no correct answers] to 3 [all correct answers]. Out of 21,177 respondents, one in ten (10.4%) did not select any correct answer, whereas one in three (33.3%) selected all three correct responses. The variable has a weak yet positive and significant correlation with the years spent in education (0.09, $p < 0.001$). We replicated the two regression models (M4 and M5 in the manuscript) testing the mediation of reflexive mindset, and substituted "Climate change: own responsibility" with institutional knowledge (see Table C.2).

The results are largely the same. Institutional knowledge has a positive and significant correlation with e-privacy protection; its addition slightly reduces the strength of the direct effect of years in education on e-privacy management, in a similar fashion compared to the models in the manuscript. Also in M5_alt, where all the mediators are added, the patterns are the same as in the main models, and the standardized coefficient (beta) of Institutional knowledge also reaches up to 0.06 ($p < 0.001$).

TABLE C.1 Multilevel linear probability model of non-internet use on individual characteristics.

Variable	Null model	MA1	MA2
Fixed Effects	b	b	b
Years in education[a]		−0.024***	−0.024***
DESI[a]			−0.009***
DESI[a]*Years in education[a]			0.001***
Female		0.016***	0.016***
Age categories			
15-24		−0.074***	−0.073***
25-34		−0.012	−0.012
35-44 (Ref)			
45-54		0.049***	0.049***
55-64		0.171***	0.171***
65-74		0.288***	0.288***
75+		0.493***	0.494***
Student (full time)		0.037**	0.037**
Unemployed		0.113***	0.113***
Rural area or village (Ref.)			
Small/middle town		−0.023***	−0.023***
Large town		−0.041***	−0.041***
Constant	0.234***	0.059**	0.059***
Random effects			
var(Years in education)		0.0001***	0.0001***
var(Country)	0.009***	0.009***	0.002***
Pseudo R^2country		0%	
Covariance Years in education with Country		−0.0002***	−0.0002***
var(Individual)	0.169***	0.104***	0.104***
Pseudo R^2individual		38.4%	38.4%
Model fit			
-2 Log Likelihood	29626.3	16320.244	16281.83
Observations	27.885	27.885	27.885

[a] Centered variable
b = coefficient; var = Variance
+ $p < 0.10$; * $p < 0.05$; ** $p < 0.01$; *** $p < 0.001$

TABLE C.2 Multilevel linear regression analyses of e-privacy management on individual characteristics.

Variables Fixed effects	M4_alt b	M5_alt b
Years in education[a]	0.06***	0.04***
Internet use index[a]		0.30***
Digital skills[a]		0.51***
Institutional knowledge	0.18***	0.13***
Control variables omitted from output		
Constant	1.99***	2.00***
Random effects		
var(Country)	0.36***	0.27***
Pseudo R^2country	1.88%	27.2%
var(Individual)	3.09***	2.89***
Pseudo R^2individual	4.2%	10.4%
Model fit		
LR chi2	183.3***b	1598.7***b
Δdf	1	3
-2 Log Likelihood	84143.5	82728.1
Observations	21177	21177

[a] Centered variable; [b] Nested in M1
b = coefficient; var = Variance; LR = Likelihood Ratio
+ p< 0.10; * $p < 0.05$; ** $p < 0.01$; *** $p < 0.001$

Appendix to Switch on the Big Brother!

Investigating the educational gradients in acceptance of online and public surveillance among European citizens

List of countries included in EVS 2017-2020: Albania (AL); Armenia (AM); Austria (AT); Azerbaijan (AZ); Bosnia and Herzegovina (BA); Belarus (BY); Bulgaria (BG); Croatia (HR); Czechia (CZ); Denmark (DK); Estonia (EE); Finland (FI); France (FR); Georgia (GE); Germany (DE); Great Britain (GB); Hungary (HU); Iceland (IS); Italy (IT); Lithuania (LT); Montenegro (ME); Netherlands (NL); North Macedonia (MK); Norway (NO); Poland (PL); Portugal (PT); Romania (RO); Russia (RU); Serbia (RS); Slovakia (SK); Slovenia (SI); Spain (ES); Sweden (SE); Switzerland (CH).

Institutional knowledge. A principal axis factor analysis with promax rotation was conducted to ascertain that the three items selected to test knowledge of democratic systems and the three items selected for the knowledge of authoritarian systems load on different factors. As showed in table D.3 the three 'democratic' items load on a factor explaining 24% of the variance, whereas the three 'authoritarian' items load on a second factor explaining 16% of the variance.

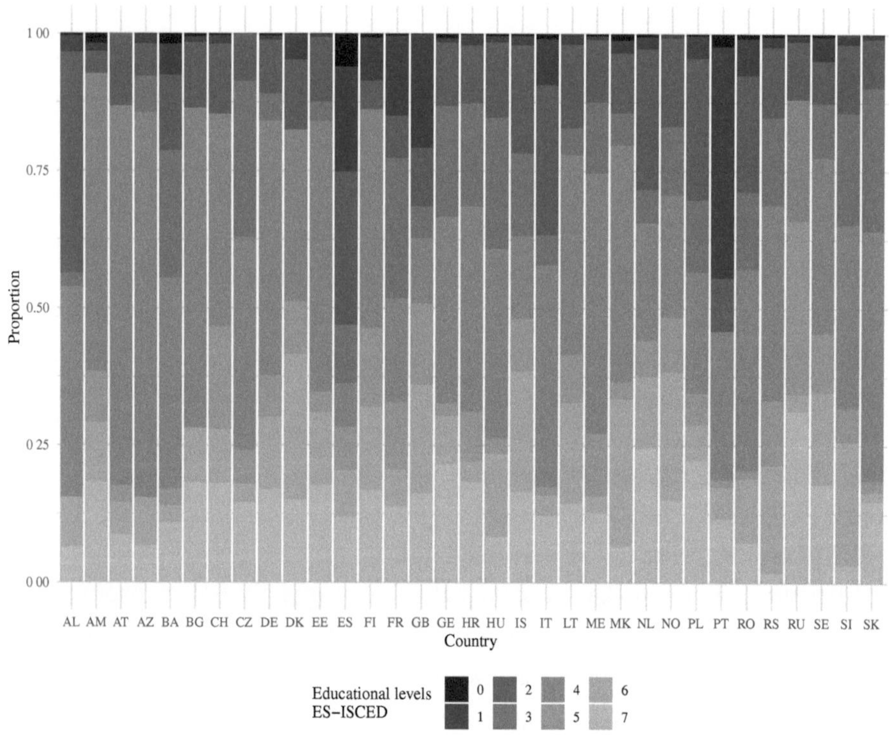

FIGURE D.1 Distribution of the ES-ISCED educational attainment by countries (N = 48,047) Source: EVS (2020), own calculations.

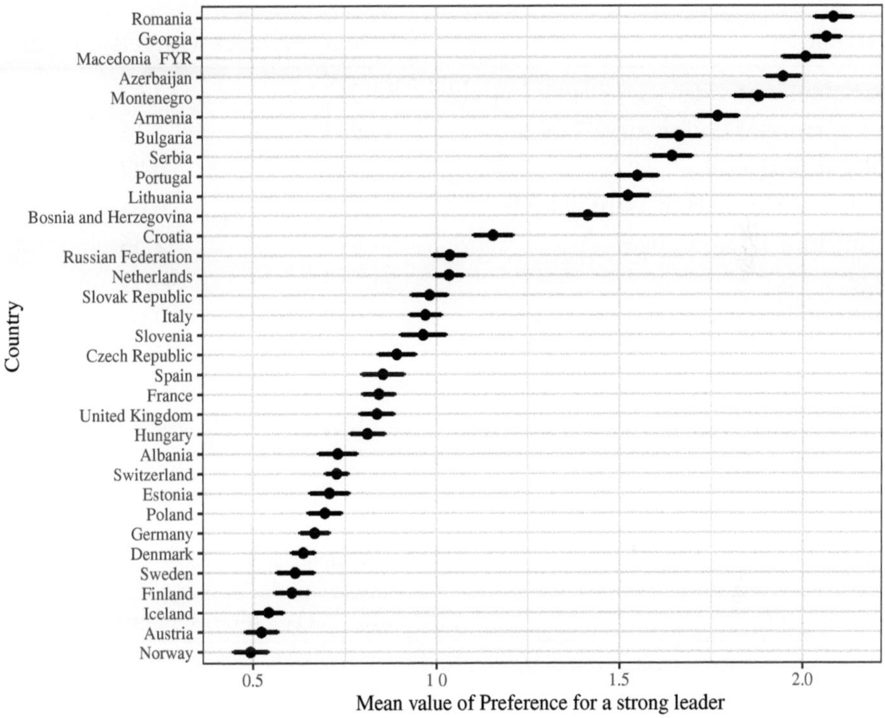

FIGURE D.2 Average score on item measuring the preference for a strong leader across countries with 95% confidence intervals. Source: EVS (2020), own calculations.

TABLE D.1 Percentage of missing values (Don't know + I prefer not to answer) per variable.

Variable	N	% cases with missing values
Acceptance of video surveillance in public	54,943	2.57
Acceptance of online surveillance	54,943	4.15
Educational level	54,943	0.82
Preference for having a strong leader	54,943	6.64
Institutional knowledge	54,943	2.36
Age	54,943	0.59
Female	54,943	0.05

TABLE D.2 Descriptive statistics of individual-level variables.

Statistic	N	Min	Max	Mean/Proportion	St. Dev.
Acceptance of video surveillance in public	48,047	1	4	2.71	1.08
Acceptance of online surveillance	48,047	1	4	1.86	0.95
Educational level	48,047	0	7	4.21	1.75
Preference for having a strong leader	48,047	0	3	1.07	1.03
Institutional knowledge	48,047	0.00	1.00	0.52	0.24
Age	48,047	18	82	49.26	17.51
Female	48,047	0	1	0.54	
Mode: Cawi	48,047	0	1	0.12	
Mode: Mail	48,047	0	1	0.03	

TABLE D.3 Factor loadings after promax rotation (N = 54,689).

Variable	Democracy	Authoritarian regime
People choose leaders	0.705	
Civil rights protect from oppression	0.651	
Women have the same rights as men	0.693	
Religious authorities interpret the law		0.585
The army takes over		0.602
People obey their rulers		0.499
Explained variance	24%	16%

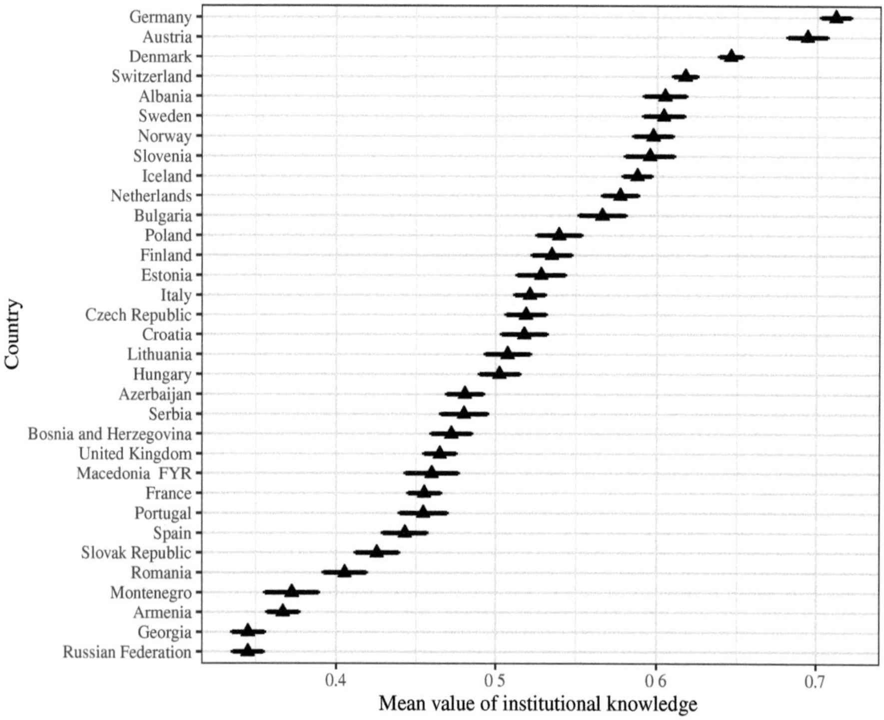

FIGURE D.3 Average score of institutional knowledge across countries with 95% confidence intervals. Source: EVS (2020), own calculations.

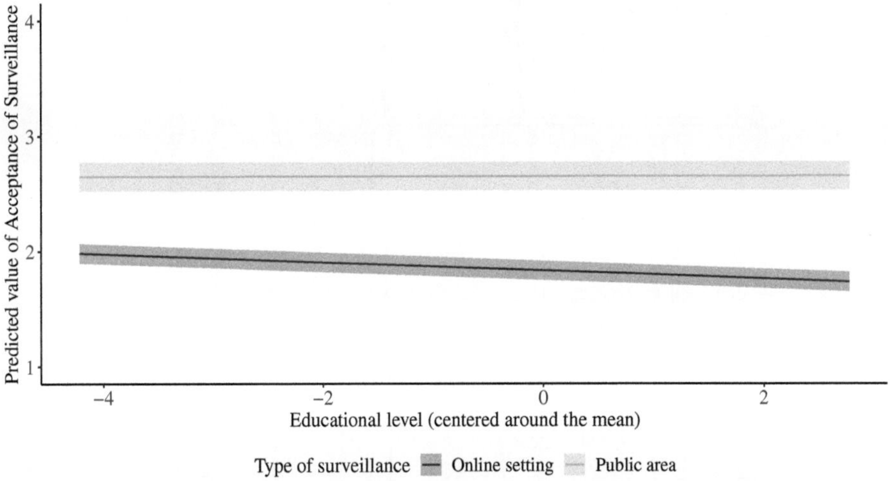

FIGURE D.4 Predicted Acceptance of Surveillance by education (based on M1 in Table 5.1). Source: EVS (2020). Own calculations

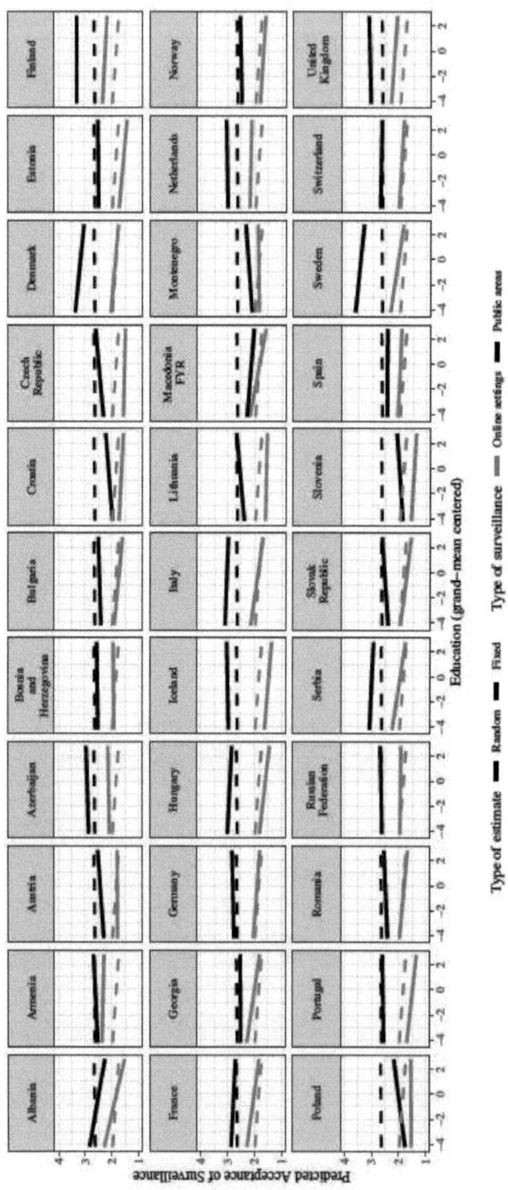

FIGURE D.5 Fixed and random slopes (estimated from M5, Table 5.2). Dashed lines represent the slopes of education for the pooled sample whereas the solid lines represent the slopes for each country. Source: EVS (2020). Own calculations.

Bibliography

Acemoglu, D. (2002). Technical change, inequality, and the labor market. *Journal of economic literature*, *40*(1), 7–72. https://www.jstor.org/stable/2698593

Achterberg, P., de Koster, W., & van der Waal, J. (2017). A science confidence gap: Education, trust in scientific methods, and trust in scientific institutions in the United States, 2014. *Public Understanding of Science*, *26*(6), 704–720. https://doi.org/10.1177/0963662515617367

Acquisti, A., Brandimarte, L., & Loewenstein, G. (2015). Privacy and human behavior in the age of information. *Science*, *347*(6221), 509–514. https://doi.org/10.1126/science.aaa1465

Ada Lovelace Institute. (2021). *What place should COVID-19 vaccine passports have in society?* (Tech. rep. February). Ada Lovelace Institute. https://www.adalovelaceinstitute.org/summary/covid-19-vaccine-passports/

Altmann, S., Milsom, L., Zillessen, H., Blasone, R., Gerdon, F., Bach, R., Kreuter, F., Nosenzo, D., Toussaert, S., & Abeler, J. (2020). Acceptability of app-based contact tracing for COVID-19: Cross-country survey study. *JMIR mHealth and uHealth*, *8*(8). https://doi.org/10.2196/19857

Andrejevic, M. (2014). The big data divide. *International Journal of Communication*, *8*, 1673–1689. https://ijoc.org/index.php/ijoc/article/view/2161/1163

Anthony, D., Campos-Castillo, C., & Horne, C. (2017). Toward a Sociology

of Privacy. *Annual Review of Sociology*, *43*(1), 249–269. https://doi.org/10.1146/annurev-soc-060116-053643
Apthorpe, N., Shvartzshnaider, Y., Mathur, A., Reisman, D., & Feamster, N. (2018). Discovering smart home internet of things privacy norms using contextual integrity. *Proceedings of the ACM on Interactive, Mobile, Wearable and Ubiquitous Technologies*, *2*(2), 1–23. https://doi.org/10.1145/3214262
Araujo, T., Ausloos, J., van Atteveldt, W., Loecherbach, F., Moeller, J., Ohme, J., Trilling, D., van de Velde, B., de Vreese, C., & Welbers, K. (2022). Osd2f: An open-source data donation framework. *Computational Communication Research*, *4*(2), 372–387. https://doi.org/10.5117/CCR2022.2.001.ARAU
Ayaburi, E. W., & Treku, D. N. (2020). Effect of penitence on social media trust and privacy concerns: The case of Facebook. *International Journal of Information Management*, *50*, 171–181. https://doi.org/10.1016/j.ijinfomgt.2019.05.014
Baller, S., Dutta, S., & Lanvin, B. (2016). *The Global Information Technology Report 2016 Innovating in the Digital Economy* (tech. rep.). https://www.weforum.org/reports/the-global-information-technology-report-2016
Bartsch, M., & Dienlin, T. (2016). Control your Facebook: An analysis of online privacy literacy. *Computers in Human Behavior*, *56*, 147–154. https://doi.org/10.1016/j.chb.2015.11.022
Beck, U. (1992). *Risk society: Towards a new modernity*. Sage.
Beck, U. (2002). The Cosmopolitan Society and Its Enemies. *Theory, Culture & Society*, *19*(2), 17–44. https://doi.org/10.1177/026327640201900101
Beck, U. (2013, August 30). The digital freedom risk: too fragile an acknowledgment. https://www.opendemocracy.net/en/can-europe-make-it/digital-freedom-risk-too-fragile-acknowledgment/
Beck, U., Bonss, W., & Lau, C. (2003). The Theory of Reflexive Modernization: Problematic, Hypotheses and Research Programme. *Theory, Culture & Society*, *20*(2), 1–33. https://doi.org/10.1177/0263276403020002001

Beer, D. (2017). The social power of algorithms. *Information, Communication & Society*, *20*(1), 1–13. https://doi.org/10.1080/1369118X.2016.1216147

Blank, G., & Dutton, W. H. (2012). Age and trust in the internet: The centrality of experience and attitudes toward technology in britain. *Social Science Computer Review*, *30*(2), 135–151. https://doi.org/10.1177/0894439310396186

Blank, G., Bolsover, G., & Dubois, E. (2014). A New Privacy Paradox: Young people and privacy on social network sites. *Global Cyber Security Capacity Centre: Draft Working Paper*. https://doi.org/10.2139/ssrn.2479938

Boeschoten, L., Ausloos, J., Möller, J. E., Araujo, T., & Oberski, D. L. (2022). A framework for privacy preserving digital trace data collection through data donation. *Computational Communication Research*, *4*(2), 388–423. https://doi.org/10.5117/CCR2022.2.002.BOES

Bol, T., & van de Werfhorst, H. G. (2011). Signals and closure by degrees: The education effect across 15 European countries. *Research in Social Stratification and Mobility*, *29*(1), 119–132. https://doi.org/10.1016/j.rssm.2010.12.002

Bonfadelli, H. (2002). The Internet and Knowledge Gaps: A Theoretical and Empirical Investigation. *European Journal of Communication*, *17*(1), 65–84. https://doi.org/10.1177/0267323102017001607

Bonoli, G. (2007). Time matters. Postindustrialization, New Social Risks, and Welfare State Adaptation in Advanced Industrial Democracies. *Comparative Political Studies*, *40*(5), 495–520. https://doi.org/10.1177/0010414005285755

Boukes, M. (2019). Agenda-setting with satire: How political satire increased TTIP's saliency on the public, media, and political agenda. *Political Communication*, *36*(3), 426–451. https://doi.org/10.1080/10584609.2018.1498816

Bourgeois, A., Birch, P., & Davydovskaia, O. (2019). *Digital Education at School in Europe*, European Commission. https://doi.org/10.2797/339457

Bovens, M., & Wille, A. (2008). Deciphering the Dutch drop: Ten explanations for decreasing political trust in The Netherlands. *International Review of Administrative Sciences, 74*(2), 283–305. https://doi.org/10.1177/0020852308091135

Bovens, M., & Wille, A. (2017). *Diploma democracy: The rise of political meritocracy*. Oxford University Press.

Boyd, D., & Crawford, K. (2012). Critical questions for big data: Provocations for a cultural, technological, and scholarly phenomenon. *Information, Communication & Society, 15*(5), 662–679. https://doi.org/10.1080/1369118X.2012.678878

Boyd, D., & Hargittai, E. (2010). Facebook privacy settings: Who cares? *First Monday, 15*(8), 1–11. https://doi.org/10.5210/fm.v15i8.3086

Bradford, B., Yesberg, J. A., Jackson, J., & Dawson, P. (2020). Live facial recognition: Trust and legitimacy as predictors of public support for police use of new technology. *The British Journal of Criminology, 60*(6), 1502–1522. https://doi.org/10.1093/bjc/azaa032

Brayne, S. (2017). Big Data Surveillance: The Case of Policing. *American Sociological Review, 82*(5), 977–1008. https://doi.org/10.1177/0003122417725865

Brown, A. J. (2020). "Should I Stay or Should I Leave?": Exploring (Dis)continued Facebook Use After the Cambridge Analytica Scandal. *Social Media + Society, 6*(1), 1–8. https://doi.org/10.1177/2056305120913884

Büchi, M., Just, N., & Latzer, M. (2017). Caring is not enough: the importance of Internet skills for online privacy protection. *Information, Communication & Society, 20*(8), 1261–1278. https://doi.org/10.1080/1369118X.2016.1229001

Buder, F., Dieckmann, A., Manewitsch, V., Dietrich, H., Wiertz, C., Banerjee, A., Acar, O. A., & Ghosh, A. (2020). *Adoption rates for contact tracing app configurations in Germany* (tech. rep.). Nuremberg Institute for Market

Decisions. https://www.nim.org/en/research/research-reports/adoption-rates-contact-tracing-app
Burgess, A., Wardman, J., & Mythen, G. (2018). Considering risk: placing the work of Ulrich Beck in context. *Journal of Risk Research*, *21*(1), 1–5. https://doi.org/10.1080/13669877.2017.1383075
Burrell, J., & Fourcade, M. (2020). The society of algorithms. *Annual Review of Sociology*, *47*, 213–237. https://doi.org/10.1146/annurev-soc-090820-020800
Calvo, R. A., Deterding, S., & Ryan, R. M. (2020). Health surveillance during COVID-19 pandemic. *The BMJ*, *369*(April). https://doi.org/10.1136/bmj.m1373
Catterberg, G., & Moreno, A. (2005). The Individual Bases of Political Trust: Trends in New and Established Democracies. *International Journal of Public Opinion Research*, *18*(1), 31–48. https://doi.org/10.1093/ijpor/edh081
Cho, H., Rivera-Sanchez, M., & Lim, S. S. (2009). A multinational study on online privacy: global concerns and local responses. *New Media & Society*, *11*(3), 395–416. https://doi.org/10.1177/1461444808101618
Cho, H., & Larose, R. (1999). Privacy Issues in Internet Surveys. *Social Science Computer Review*, *17*(4), 421–434. https://doi.org/10.1177/089443939901700402
Christensen, R. H. B. (2019). Ordinal—regression models for ordinal data [R package version 2019.12-10]. https://CRAN.R-project.org/package=ordinal
CROss-National Online Survey panel. (2018). CRONOS Wave 1 [CRONOS_Wave1_e01.sav]. *NSD - Norwegian Centre for Research Data, Norway – Data Archive and distributor of CRONOS data for ESS ERIC.*
Croteau, D., & Hoynes, W. (2019). *Media/society: Technology, industries, content and users* (6th). Sage.
Cruz-Jesus, F., Vicente, M. R., Bacao, F., & Oliveira, T. (2016). The education-

related digital divide: An analysis for the EU-28. *Computers in Human Behavior*, *56*, 72–82. https://doi.org/10.1016/j.chb.2015.11.027

Cukier, K., & Mayer-Schoenberger, V. (2013). The Rise of Big Data. How it's changing the Way We Think About the World. *Foreign Affairs*, *92*(3), 28–40. https://www.jstor.org/stable/23526834

Curran, D. (2013). Risk society and the distribution of bads: Theorizing class in the risk society. *British Journal of Sociology*, *64*(1), 44–62. https://doi.org/10.1111/1468-4446.12004

Davis, D. W., & Silver, B. D. (2004). Civil Liberties vs. Security: Public Opinion in the Context of the Terrorist Attacks on America. *American Journal of Political Science*, *48*(1), 28–46. https://doi.org/10.2307/1519895

De Keere, K. (2010). Wantrouwen in wetenschap: een kwestie van reflexiviteit of maatschappelijk onbehagen? *Sociologie*, *6*(1), 26–45. https://edu.nl/4emt4

de Koster, W., van der Waal, J., Achterberg, P., & Houtman, D. (2008). The Rise of the Penal State. Neo-Liberalization or New Political Culture? *British Journal of Sociology*, *48*(6), 720–734. https://doi.org/10.1093/bjc/azn057

Debatin, B., Lovejoy, J. P., Horn, A. K., & Hughes, B. N. (2009). Facebook and online privacy: Attitudes, behaviors, and unintended consequences. *Journal of Computer-Mediated Communication*, *15*(1), 83–108. https://doi.org/10.1111/j.1083-6101.2009.01494.x

Degli Esposti, S., & Santiago Gómez, E. (2015). Acceptable surveillance-orientated security technologies: Insights from the surprise project. *Surveillance and Society*, *13*(3-4), 437–454. https://doi.org/10.24908/ss.v13i3/4.5400

Dohle, S., Wingen, T., & Schreiber, M. (2020). Acceptance and adoption of protective measures during the COVID-19 pandemic: The role of trust in politics and trust in science. *Social Psychological Bulletin*, *15*(4), 1–23. https://doi.org/10.32872/spb.4315

Douglas, M., & Wildavsky, A. (1983). *Risk and Culture: An Essay on the Selec-*

tion of Technological and Environmental Dangers. University of California Press. https://www.jstor.org/stable/10.1525/j.ctt7zw3mr

Dreher, A. (2006). Does globalization affect growth? Evidence from a new index of globalization. *Applied Economics*, *38*(10), 1091–1110. https://doi.org/10.1080/00036840500392078

Dubois, E., & Blank, G. (2018). The echo chamber is overstated: The moderating effect of political interest and diverse media. *Information, Communication & Society*, *21*(5), 729–745. https://doi.org/10.1080/1369118X.2018.1428656

Dutton, W. H., & Shepherd, A. (2006). Trust in the internet as an experience technology. *Information, Communication & Society*, *9*(4), 433–451. https://doi.org/10.1080/13691180600858606

Elliott, A. (2002). Beck's sociology of risk: A critical assessment. *Sociology*, *36*(2), 293–315. https://doi.org/10.1177/0038038502036002004

Eubanks, V. (2018). *Automating inequality : how high-tech tools profile, police, and punish the poor*. St. Martin's Press.

European Commission. (2019). *2nd Survey of Schools: ICT in Education Objective 1: Benchmark progress in ICT in schools* (tech. rep.). Publications Office of the European Union. Luxembourg. https://doi.org/10.2759/23401

European Commission and European Parliament. (2017). Eurobarometer 87.1. https://doi.org/10.4232/1.12922

European Union. (2019). Special Eurobarometer 487a: The General Data Protection Regulation. https://doi.org/10.2838/579882

European Values Study. (2020). Integrated Dataset (EVS 2017). *GESIS Data Archive*. https://doi.org/10.4232/1.13560

Eurostat. (2020). Population by educational attainment level, sex and age (%). https://ec.europa.eu/eurostat/databrowser/view/EDAT_LFSE_03__custom_665742/default/table?lang=en

Fisher, J., Van Heerde, J., & Tucker, A. (2010). Does one trust judgement fit all? linking theory and empirics. *The British Journal of Politics and International*

Relations, *12*(2), 161–188. https://doi.org/10.1111/j.1467-856X.2009.00401.x

Flanagan, S. C. (1987). Value Change in Industrial Societies: Reply to Inglehart. *The American Political Science Review*, *81*(4), 1303–1319. https://doi.org/10.2307/1962590

Gerdon, F., Nissenbaum, H., Bach, R. L., Kreuter, F., & Zins, S. (2021). Individual Acceptance of Using Health Data for Private and Public Benefit: Changes During the COVID-19 Pandemic. *Harvard Data Science Review*, *Special Issue 1*. https://doi.org/10.1162/99608f92.edf2fc97

Gerosa, T., Gui, M., Hargittai, E., & Nguyen, M. H. A. O. (2021). (Mis) informed During COVID-19: How Education Level and Information Sources Contribute to Knowledge Gaps. *International Journal of Communication*, *15*, 2196–2217. https://ijoc.org/index.php/ijoc/article/view/16438

Giddens, A. (1990). *The Consequences of Modernity*. Polity.

Giddens, A. (1999). Risk and Responsibility. *The Modern Law Review*, *62*(1), 1–10. https://doi.org/10.1111/1468-2230.00188

Gilbert, A. S. (2018). Algorithmic culture and the colonization of life-worlds. *Thesis Eleven*, *146*(1), 87–96. https://doi.org/10.1177/0725513618776699

Gillespie, T. (2010). The politics of 'platforms'. *New Media & Society*, *12*(3), 347–364. https://doi.org/10.1177/1461444809342738

Gillespie, T. (2014). The relevance of algorithms. In T. Gillespie, P. Boczkowski, & K. Foot (Eds.), *Media technologies: Essays on communication, materiality, and society* (pp. 167–194). MIT Press.

Gurinskaya, A. (2020). Predicting citizens' support for surveillance cameras. Does police legitimacy matter? *International Journal of Comparative and Applied Criminal Justice*, *44*(1-2), 63–83. https://doi.org/10.1080/01924036.2020.1744027

Gygli, S., Haelg, F., Potrafke, N., & Sturm, J. E. (2019). The KOF Globalisation Index – revisited. *Review of International Organizations*, *14*(3), 543–574. https://doi.org/10.1007/s11558-019-09344-2

Hakhverdian, A., & Mayne, Q. (2012). Institutional trust, education, and corruption: A micro-macro interactive approach. *Journal of Politics*, *74*(3), 739–750. https://doi.org/10.1017/S0022381612000412

Hall, K. (2020). Public penitence: Facebook and the performance of apology. *Social Media + Society*, *6*(2), 1–10. https : / / doi . org / 10 . 1177 / 2056305120907945

Hallam, C., & Zanella, G. (2017). Online self-disclosure: The privacy paradox explained as a temporally discounted balance between concerns and rewards. *Computers in Human Behavior*, *68*, 217–227. https://doi.org/10.1016/j.chb.2016.11.033

Hanitzsch, T., Van Dalen, A., & Steindl, N. (2018). Caught in the nexus: A comparative and longitudinal analysis of public trust in the press. *International Journal of Press/Politics*, *23*(1), 3–23. https://doi.org/10.1177/1940161217740695

Hardin, R. (1993). The street-level epistemology of trust. *Politics & Society*, *21*(4), 505–529. https://doi.org/10.1177/0032329293021004006

Hardin, R. (2006). *Trust*. Polity Press.

Hargittai, E. (2002). Second-Level Digital Divide : Differences in People's Online Skills. *First monday*, *7*(4), 1–19. https://doi.org/10.5210/fm.v7i4.942

Hargittai, E., Gruber, J., Djukaric, T., Fuchs, J., & Brombach, L. (2020). Black box measures? how to study people's algorithm skills. *Information, Communication & Society*, *23*(5), 764–775. https://doi.org/10.1080/1369118X.2020.1713846

Hargittai, E., & Marwick, A. (2016). "What Can I Really Do?" Explaining the Privacy Paradox with Online Apathy. *International Journal of Communication*, *10*, 3737–3757. https://ijoc.org/index.php/ijoc/article/view/4655

Hatuka, T., & Toch, E. (2017). Being visible in public space: The normalisation of asymmetrical visibility. *Urban Studies*, *54*(4), 984–998. https://doi.org/10.1177/0042098015624384

Helsper, E. J., & Reisdorf, B. C. (2017). The emergence of a "digital underclass" in Great Britain and Sweden: Changing reasons for digital exclusion. *New Media & Society*, *19*(8), 1253–1270. https://doi.org/10.1177/1461444816634676

Hendriks, F., Kienhues, D., & Bromme, R. (2016). Trust in Science and the Science of Trust. In B. Blöbaum (Ed.), *Trust and communication in a digitized world: Models and concepts of trust research* (pp. 143–159). Springer International Publishing. https://doi.org/10.1007/978-3-319-28059-2_8

Hetherington, M. J. (1998). The political relevance of political trust. *American political science review*, *92*(4), 791–808. https://doi.org/10.2307/2586304

Hooghe, M. (2011). Why there is basically only one form of political trust. *The British Journal of Politics and International Relations*, *13*(2), 269–275. https://doi.org/10.1111/j.1467-856X.2010.00447.x

Horne, C., Darras, B., Bean, E., Srivastava, A., & Frickel, S. (2015). Privacy, technology, and norms: The case of Smart Meters. *Social Science Research*, *51*, 64–76. https://doi.org/10.1016/j.ssresearch.2014.12.003

Horne, C., & Przepiorka, W. (2019). Technology use and norm change in online privacy: experimental evidence from vignette studies. *Information, Communication & Society*, *24*(9), 1212–1228. https://doi.org/10.1080/1369118X.2019.1684542

Hox, J. J. (2002). *Multilevel analysis: techniques and applications*. Lawrence Erlbaum Associates.

Ignazi, P. (1992). The silent counter-revolution: Hypotheses on the emergence of extreme right-wing parties in Europe. *European Journal of Political Research*, *22*(1), 3–34. https://doi.org/10.1111/j.1475-6765.1992.tb00303.x

Inglehart, R. (1977). *Silent revolution: Changing values and political styles among western publics*. Princeton University Press.

Inglehart, R. (1981). Post-materialism in an environment of insecurity. *The American Political Science Review*, *75*(4), 880–900. https://doi.org/10.2307/1962290

Inglehart, R. (1997). *Modernization and postmodernization: Cultural, economic, and political change in 43 societies*. Princeton University Press.

Ioannou, A., & Tussyadiah, I. (2021). Privacy and surveillance attitudes during health crises : Acceptance of surveillance and privacy protection behaviours. *Technology in Society*, *67*, 1–16. https://doi.org/10.1016/j.techsoc.2021.101774

Jordan, J. J., Yoeli, E., & Rand, D. G. (2021). Don't get it or don't spread it: comparing self-interested versus prosocial motivations for COVID-19 prevention behaviors. *Scientific Reports*, *11*(20222), 1–17. https://doi.org/10.1038/s41598-021-97617-5

Joyce, K., Smith-Doerr, L., Alegria, S., Bell, S., Cruz, T., Hoffman, S. G., Noble, S. U., & Shestakofsky, B. (2021). Toward a Sociology of Artificial Intelligence: A Call for Research on Inequalities and Structural Change. *Socius*, *7*, 1–11. https://doi.org/10.1177/2378023121999581

Kidd, D., & McIntosh, K. (2016). Social media and social movements. *Sociology Compass*, *10*(9), 785–794. https://doi.org/10.1111/soc4.12399

Kitchin, R. (2014). Big data, new epistemologies and paradigm shifts. *Big data & society*, *1*(1), 1–12. https://doi.org/10.1177/2053951714528481

Knight, A. J., & Warland, R. (2005). Determinants of food safety risks: A multi-disciplinary approach. *Rural Sociology*, *70*(2), 253–275. https://doi.org/10.1526/0036011054776389

Kokkoris, M. D., & Kamleitner, B. (2020). Would You Sacrifice Your Privacy to Protect Public Health? Prosocial Responsibility in a Pandemic Paves the Way for Digital Surveillance. *Frontiers in Psychology*, *11*, 1–8. https://doi.org/10.3389/fpsyg.2020.578618

Kokolakis, S. (2017). Privacy attitudes and privacy behaviour: A review of current research on the privacy paradox phenomenon. *Computers & Security*, *64*, 122–134. https://doi.org/10.1016/j.cose.2015.07.002

Kolenikov, S., & Angeles, G. (2004). *The Use of Discrete Data in PCA: Theory, Simulations, and Applications to Socioeconomic Indices*, CPC/MEASURE.

https://www.measureevaluation.org/resources/publications/wp-04-85.html

Lankton, N. K., & McKnight, D. H. (2011). What does it mean to trust facebook? examining technology and interpersonal trust beliefs. *SIGMIS Database*, *42*(2), 32–54. https://doi.org/10.1145/1989098.1989101

Latour, B. (2003). Is Re-modernization Occurring - And If So, How to Prove It?: A Commentary on Ulrich Beck. *Theory, Culture & Society*, *20*(2), 35–48. https://doi.org/10.1177/0263276403020002002

Levi, M., & Stoker, L. (2000). Political trust and trustworthiness. *Annual review of political science*, *3*, 475–507. https://doi.org/10.1146/annurev.polisci.3.1.475

Lewandowsky, S., Dennis, S., Perfors, A., Kashima, Y., White, J. P., Garrett, P., Little, D. R., & Yesilada, M. (2021). Public acceptance of privacy-encroaching policies to address the COVID-19 pandemic in the United Kingdom. *PLoS ONE*, *16*(1), 1–23. https://doi.org/10.1371/journal.pone.0245740

Lind, F., & Boomgaarden, H. G. (2019). What we do and don't know: a meta-analysis of the knowledge gap hypothesis. *Annals of the International Communication Association*, *43*(3), 210–224. https://doi.org/10.1080/23808985.2019.1614475

Litt, E. (2013a). Measuring users' internet skills: A review of past assessments and a look toward the future. *New Media & Society*, *15*(4), 612–630. https://doi.org/10.1177/1461444813475424

Litt, E. (2013b). Understanding social network site users' privacy tool use. *Computers in Human Behavior*, *29*(4), 1649–1656. https://doi.org/10.1016/j.chb.2013.01.049

Litt, E., & Hargittai, E. (2014). A bumpy ride on the information superhighway: Exploring turbulence online. *Computers in Human Behavior*, *36*, 520–529. https://doi.org/10.1016/j.chb.2014.04.027

Luijkx, R., Jonsdottir, G., Gummer, T., Ernst Staehli, M., Frederiksen, M.,

Ketola, K., Reeskens, T., Brislinger, E., Christmann, P., Gunnarsson, S., Bragi Hjaltason, A., Joye, D., Lomazzi, V., Maineri, A., Milbert, P., Ochsner, M., Pollien, A., Sapin, M., Solanes, I., ... Wolf, C. (2021). The European Values Study 2017: On the way to the future using mixed-modes. *European Sociological Review*, *37*(2), 330–346. https://doi.org/10.1093/esr/jcaa049

Lupton, D. (1997). Consumerism, reflexivity and the medical encounter. *Social Science & Medicine*, *45*(3), 373–381. https://doi.org/10.1016/S0277-9536(96)00353-X

Lupton, D. (2016). Digital Risk Society. In A. Burgess, A. Alemanno, & J. Zinn (Eds.), *The routledge handbook of risk studies* (pp. 301–309). Routledge. https://doi.org/10.2139/ssrn.2511717

Lupton, D., & Michael, M. (2017). 'Depends on who's got the data': Public understandings of personal digital dataveillance. *Surveillance & Society*, *15*(2), 254–268. https://doi.org/10.24908/ss.v15i2.6332

Lutz, C. (2019). Digital inequalities in the age of artificial intelligence and big data. *Human Behavior and Emerging Technologies*, *1*(2), 141–148. https://doi.org/10.1002/hbe2.140

Lyon, D. (2005). Surveillance as social sorting. In D. Lyon (Ed.), *Surveillance as social sorting. privacy, risk, and digital discrimination* (pp. 13–30). Routledge.

Lyon, D. (2014). Surveillance, Snowden, and Big Data: Capacities, consequences, critique. *Big Data & Society*, *1*(2), 1–13. https://doi.org/10.1177/2053951714541861

Makarovs, K., & Achterberg, P. (2017). Contextualizing educational differences in "vaccination uptake": A thirty nation survey. *Social Science and Medicine*, *188*, 1–10. https://doi.org/10.1016/j.socscimed.2017.06.039

Mann, M., & Matzner, T. (2019). Challenging algorithmic profiling: The limits of data protection and anti-discrimination in responding to emergent discrimination. *Big Data & Society*, *6*(2), 1–11. https://doi.org/10.1177/2053951719895805

Manovich, L. (2011). Trending: The promises and the challenges of big social data. In M. K. Gold (Ed.), *Debates in the digital humanities* (pp. 460–475). The University of Minnesota Press. https://doi.org/10.5749/minnesota/9780816677948.003.0047

Marwick, A. E., & Hargittai, E. (2018). Nothing to hide, nothing to lose? Incentives and disincentives to sharing information with institutions online. *Information, Communication & Society, 22*(12), 1697–1713. https://doi.org/10.1080/1369118X.2018.1450432

Marx, G. T. (2016). *Windows into the Soul. Surveillance and Society in an Age of High Technology*. The University of Chicago Press.

Mascheroni, G. (2020). Datafied childhoods: Contextualising datafication in everyday life. *Current Sociology, 68*(6), 798–813. https://doi.org/10.1177/0011392118807534

Massicotte, P., & Eddelbuettel, D. (2021). *Gtrendsr: Perform and display google trends queries* [R package version 1.5.0]. https://CRAN.R-project.org/package=gtrendsR

Mayer, R. C., Davis, J. H., & Schoorman, F. D. (1995). An integrative model of organizational trust. *Academy of Management Review, 20*(3), 709–734. https://doi.org/10.5465/AMR.1995.9508080335

McCarthy, M. T. (2016). The big data divide and its consequences. *Sociology Compass, 10*(12), 1131–1140. https://doi.org/10.1111/soc4.12436

McIlroy, D., Brownrigg, R., Minka, T. P., & Bivand, R. (2018). mapproj: Map Projections.

Mejias, U. A., & Couldry, N. (2019). Datafication. *Internet Policy Review, 8*(4), 1–10. https://doi.org/10.14763/2019.4.1428

Mennicken, A., & Espeland, W. N. (2019). What's New with Numbers? Sociological Approaches to the Study of Quantification. *Annual Review of Sociology, 45*, 223–245. https://doi.org/10.1146/annurev-soc-073117-041343

Merchant, B. (2018). Predictim Claims Its AI Can Flag 'Risky' Babysitters.

So I Tried It on the People Who Watch My Kids. Retrieved August 27, 2021, from https://gizmodo.com/predictim-claims-its-ai-can-flag-risky-babysitters-so-1830913997

Meyer, S., Ward, P., Coveney, J., & Rogers, W. (2008). Trust in the health system- An analysis and extension of the social theories of Giddens and Luhmann. *Health Sociology Review*, *17*(2), 177–186. https://doi.org/10.5172/hesr.451.17.2.177

Mishler, W., & Rose, R. (2001). What are the origins of political trust? testing institutional and cultural theories in post- communist societies. *Comparative Political Studies*, *34*(1), 30–62. https://doi.org/10.1177/0010414001034001002

Mols, A., & Janssen, S. (2017). Not Interesting Enough to be Followed by the NSA: An analysis of Dutch privacy attitudes. *Digital Journalism*, *5*(3), 277–298. https://doi.org/10.1080/21670811.2016.1234938

Mythen, G. (2004). *Ulrich Beck: a critical introduction to the risk society.*

Nettleton, S., & Burrows, R. (2003). ICTs and processes of reflexive modernization. *Critical Social Policy*, *23*(2), 165–185. https://doi.org/10.1177/02610183030230002003

Neuman, W. R., Bimber, B., & Hindman, M. (2011). The Internet and Four Dimensions of Citizenship. In R. Shapiro & R. Jacob (Eds.), *The Oxford Handbook of American Public Opinion and the Media* (pp. 22–42). Oxford University Press. https://doi.org/10.1093/oxfordhb/9780199545636.003.0002

Newlands, G., Lutz, C., Tamò-Larrieux, A., Villaronga, E. F., Harasgama, R., & Scheitlin, G. (2020). Innovation under pressure: Implications for data privacy during the Covid-19 pandemic. *Big Data & Society*, *7*(2), 1–14. https://doi.org/10.1177/2053951720976680

Newton, K., & Zmerli, S. (2011). Three forms of trust and their association. *European Political Science Review*, *3*(2), 169–200. https://doi.org/10.1017/S1755773910000330

Newton, K. (2001). Trust, social capital, civil society, and democracy. *International Political Science Review*, *22*(2), 201–214. https://doi.org/10.1177/0192512101222004

Newton, K., Stolle, D., & Zmerli, S. (2018). Social and political trust. In E. M. Uslaner (Ed.), *The oxford handbook of social and political trust*. Oxford University Press. https://doi.org/10.1093/oxfordhb/9780190274801.013.20

Nisbet, M., & Markowitz, E. M. (2014). Understanding public opinion in debates over biomedical research: Looking beyond political partisanship to focus on beliefs about science and society. *PLoS ONE*, *9*(2), 1–12. https://doi.org/10.1371/journal.pone.0088473

Nissenbaum, H. (2004). Privacy as contextual integrity. *Washington Law Review*, *79*(1), 119–158. https://digitalcommons.law.uw.edu/wlr/vol79/iss1/10

Nissenbaum, H. (2010). *Privacy In Context: Technology Policy And The Integrity Of Social Life*. Stanford University Press.

Nissenbaum, H. (2011). A Contextual Approach to Privacy Online. *Daedalus*, *140*(4), 32–48. https://ssrn.com/abstract=2567042

Nissenbaum, H. (2019). Contextual integrity up and down the data food chain. *Theoretical Inquiries in Law*, *20*(1), 221–256. https://doi.org/10.1515/til-2019-0008

Norris, P. (2011). *Democratic deficit: Critical citizens revisited*. Cambridge University Press. https://doi.org/10.1017/CBO9780511973383

Norris, P., & Inglehart, R. (2019). *Cultural Backlash Trump, Brexit, and the Rise of Authoritarian Populism*. Cambridge University Press. https://doi.org/10.1017/9781108595841

O'Connell, A. (2011). *Logistic Regression Models for Ordinal Response Variables*. Sage Publications, Inc. https://doi.org/10.4135/9781412984812

Olson, J. S., Grudin, J., & Horvitz, E. (2005). A study of preferences for sharing and privacy. *Conference on Human Factors in Computing Systems - Proceedings*, 1985–1988. https://doi.org/10.1145/1056808.1057073

Oude Groeniger, J., Noordzij, K., van der Waal, J., & de Koster, W. (2021). Dutch COVID-19 lockdown measures increased trust in government and trust in science: A difference-in-differences analysis. *Social Science & Medicine*, *275*, 1–8. https://doi.org/10.1016/j.socscimed.2021.113819

Park, S., & Humphry, J. (2019). Exclusion by design: intersections of social, digital and data exclusion. *Information, Communication & Society*, *22*(7), 934–953. https://doi.org/10.1080/1369118X.2019.1606266

Park, Y. J. (2011). Digital Literacy and Privacy Behavior Online. *Communication Research*, *40*(2), 215–236. https : / / doi . org / 10 . 1177 / 0093650211418338

Park, Y. J. (2013). Offline Status, Online Status: Reproduction of Social Categories in Personal Information Skill and Knowledge. *Social Science Computer Review*, *31*(6), 680–702. https://doi.org/10.1177/0894439313485202

Park, Y. J. (2015). Do men and women differ in privacy? Gendered privacy and (in)equality in the Internet. *Computers in Human Behavior*, *50*, 252–258. https://doi.org/10.1016/j.chb.2015.04.011

Park, Y. J. (2018). Social antecedents and consequences of political privacy. *New Media & Society*, *20*(7), 2352–2369. https://doi.org/10.1177/1461444817716677

Park, Y. J., & Chung, J. E. (2017). Health privacy as sociotechnical capital. *Computers in Human Behavior*, *76*, 227–236. https://doi.org/10.1016/j.chb.2017.07.025

Park, Y. J., & Shin, D. (2020). Contextualizing privacy on health-related use of information technology. *Computers in Human Behavior*, *105*, 1–9. https://doi.org/10.1016/j.chb.2019.106204

Pavone, V., & Degli Esposti, S. (2012). Public assessment of new surveillance-oriented security technologies: Beyond the trade-off between privacy and security. *Public Understanding of Science, 21*(5), 556–572. https://doi.org/10.1177/0963662510376886

Pinheiro, J., Bates, D., DebRoy, S., Sarkar, D., & R Core Team. (2021). *nlme: Linear and nonlinear mixed effects models* [R package version 3.1-153]. https://CRAN.R-project.org/package=nlme

Price, A. M., & Peterson, L. P. (2016). Scientific progress, risk, and development: Explaining attitudes toward science cross-nationally. *International Sociology, 31*(1), 57–80. https://doi.org/10.1177/0268580915614593

R Core, T. (2013). R: A Language and Environment for Statistical Computing.

R Core Team. (2021). *R: A language and environment for statistical computing.* R Foundation for Statistical Computing. Vienna, Austria. https://www.R-project.org/

Reeskens, T., Muis, Q., Sieben, I., Vandecasteele, L., Luijkx, R., & Halman, L. (2021). Stability or change of public opinion and values during the coronavirus crisis? Exploring Dutch longitudinal panel data. *European Societies, 23*(sup1), S153–S171. https://doi.org/10.1080/14616696.2020.1821075

Robinson, L., Cotten, S. R., Ono, H., Quan-Haase, A., Mesch, G., Chen, W., Schulz, J., Hale, T. M., & Stern, M. J. (2015). Digital inequalities and why they matter. *Information, Communication & Society, 18*(5), 569–582. https://doi.org/10.1080/1369118X.2015.1012532

Rogers, E. M. (2003). *Diffusion of innovations* (5th ed.). Free Press.

Rothstein, B., & Stolle, D. (2008). The state and social capital: An institutional theory of generalized trust. *Comparative Politics, 40*(4), 441–459. https://doi.org/10.5129/001041508X12911362383354

Samatas, M. (2005). Studying surveillance in Greece: methodological and other problems related to an authoritarian surveillance culture. *Surveillance and Society, 3*(2-3), 181–197. https://doi.org/10.24908/ss.v3i2/3.3500

Scheerder, A., van Deursen, A. J., & van Dijk, J. A. (2017). Determinants of Internet skills, use and outcomes. A systematic review of the second- and third-level digital divide. https://doi.org/10.1016/j.tele.2017.07.007

Scheufele, D. A., & Tewksbury, D. (2007). Framing, agenda setting, and priming: The evolution of three media effects models. *Journal of Communication*, *57*(4), 9–20. https://doi.org/10.1111/j.0021-9916.2007.00326.x

Schlegel, B., & Steenbergen, M. (2020). *Brant: Test for parallel regression assumption* [R package version 0.3-0]. https://CRAN.R-project.org/package=brant

Schneider, S. L. (2009). *Confusing Credentials: The Cross-Nationally Comparable Measurement of Educational Attainment* (Doctoral dissertation April). https://doi.org/10.13140/RG.2.2.35997.51683

Sharp, C. (2002). School Starting Age: European Policy and Recent Research. *LGA Seminar 'When Should our Children Start School?'* https://www.nfer.ac.uk/nfer/publications/44410/44410.pdf

Snijders, T. A. B., & Bosker, R. J. (1999). Multivariate Multilevel Models. In *Multilevel analysis. an introduction to basic and advanced multilevel modeling* (pp. 200–206). SAGE Publications.

Solove, D. J. (2007). 'i've got nothing to hide' and other misunderstandings of privacy. *San Diego Law Review*, *44*, 745–772. https://ssrn.com/abstract=998565

South, A. (2011). rworldmap: A New R package for Mapping Global Data. *The R Journal*, *3*(1), 35–43.

Southerton, C. (2020). Datafication. In L. A. Schintler & C. L. McNeely (Eds.), *Encyclopedia of big data* (pp. 1–4). https://doi.org/10.1007/978-3-319-32001-4_332-1

StataCorp. (2019). Stata Statistical Software: Release 16.

Stubager, R. (2008). Education effects on authoritarian-libertarian values: A question of socialization. *British Journal of Sociology*, *59*(2), 327–350. https://doi.org/10.1111/j.1468-4446.2008.00196.x

Trüdinger, E. M., & Steckermeier, L. C. (2017). Trusting and controlling? Political trust, information and acceptance of surveillance policies: The case of Germany. *Government Information Quarterly*, *34*(3), 421–433. https://doi.org/10.1016/j.giq.2017.07.003

Tsfati, Y., & Ariely, G. (2014). Individual and contextual correlates of trust in media across 44 countries. *Communication Research*, *41*(6), 760–782. https://doi.org/10.1177/0093650213485972

Uslaner, E. M. (2002). *The moral foundations of trust*. Cambridge University Press. https://doi.org/10.1017/CBO9780511614934

van de Werfhorst, H. G., & de Graaf, N. D. (2004). The sources of political orientations in post-industrial society: Social class and education revisited. *British Journal of Sociology*, *55*(2), 211–235. https://doi.org/10.1111/j.1468-4446.2004.00016.x

van den Broek, T., Ooms, M., Friedewald, M., van Lieshout, M., & Rung, S. (2017). Privacy and security. Citizens' desires for an equal footing. In M. Friedewald, J. P. Burgess, J. Čas, R. Bellanova, & W. Peissl (Eds.), *Surveillance, privacy and security. citizens' perspective* (pp. 15–35). https://doi.org/10.4324/9781315619309

van der Meer, J. (2003). Rain or fog? an empirical examination of social capital's rainmaker effects. *Generating Social Capital*, 133–151. https://doi.org/10.1057/9781403979544_7

van Dijck, J. (2014). Datafication, dataism and dataveillance: Big Data between scientific paradigm and ideology. *Surveillance & Society*, *12*(2), 197–208. https://hdl.handle.net/11245/1.435997

van Heek, J., Arning, K., & Ziefle, M. (2017). The Surveillance Society: Which Factors Form Public Acceptance of Surveillance Technologies? In M. Helfert, C. Klein, B. Donnellan, & O. Gusikhin (Eds.), *Smart cities, green technologies, and intelligent transport systems* (pp. 170–191). Springer International Publishing. https://doi.org/10.1007/978-3-319-63712-9_10

van Deursen, A. J., & Helsper, E. J. (2015). The Third-Level Digital Divide: Who Benefits Most from Being Online? In *Communication and informa-*

tion technologies annual (pp. 29–52). https://doi.org/10.1108/S2050-206020150000010002

van Deursen, A. J., van Dijk, J. A., & Peters, O. (2011). Rethinking Internet skills: The contribution of gender, age, education, Internet experience, and hours online to medium- and content-related Internet skills. *Poetics*, *39*(2), 125–144. https://doi.org/10.1016/j.poetic.2011.02.001

van Dijck, J., Poell, T., & de Waal, M. (2016). *De platformsamenleving. strijd om publieke waarden in een online wereld*. Amsterdam: Amsterdam University Press. https://hdl.handle.net/11245/1.544830

van Dijk, J. A. (2005). *The Deepening Divide: Inequality in the Information Society*. Sage Publications. https://doi.org/10.4135/9781452229812

van Dijk, J. A. (2013). A theory of the digital divide. In M. Ragnedda & G. W. Muschert (Eds.), *The digital divide: The internet and social inequality in international perspective* (pp. 29–51). Routledge. https://doi.org/10.4324/9780203069769

van Ingen, E., & Bekkers, R. (2015). Generalized Trust Through Civic Engagement? Evidence from Five National Panel Studies. *Political Psychology*, *36*(3), 277–294. https://doi.org/10.1111/pops.12105

Venables, W. N., & Ripley, B. D. (2002). *Modern applied statistics with s* (Fourth) [ISBN 0-387-95457-0]. Springer. https://www.stats.ox.ac.uk/pub/MASS4/

Vitak, J., & Zimmer, M. (2020). More Than Just Privacy : Using Contextual Integrity to Evaluate the Long-Term Risks from COVID-19 Surveillance Technologies. *Social Media + Society*, *6*(3), 1–4. https://doi.org/10.1177/2056305120948250

Ward, P. (2006). Trust, Reflexivity and Dependence: A 'Social Systems Theory' Analysis in/of Medicine. *European Journal of Social Quality*, *6*(2), 143–158. https://doi.org/10.3167/ejsq.2006.060208

Ward, P., & Coates, A. (2006). 'We shed tears, but there is no one there to wipe them up for us': Narratives of (mis)trust in a materially deprived community. *Health, 10*(3), 283–301. https://doi.org/10.1177/1363459306064481

Wegscheider, C., & Stark, T. (2020). What drives citizens ' evaluation of democratic performance ? The interaction of citizens ' democratic. *Zeitschrift fur Vergleichende Politikwissenschaft*. https://doi.org/10.1007/s12286-020-00467-0

Wester, M., & Giesecke, J. (2019). Accepting surveillance – An increased sense of security after terror strikes? *Safety Science, 120*, 383–387. https://doi.org/10.1016/j.ssci.2019.07.013

Wickham, H. (2016). *ggplot2: Elegant Graphics for Data Analysis*. https://ggplot2.tidyverse.org

Wildavsky, A., & Dake, K. (1990). Theories of risk perception: Who fears what and why? *Daedalus, 119*(4), 41–60. http://www.jstor.org/stable/20025337

Wimmer, J., & Quandt, T. (2006). Living in the risk society: An interview with Ulrich Beck. *Journalism Studies, 7*(2), 336–347. https://doi.org/10.1080/14616700600645461

Wnuk, A., Oleksy, T., & Domaradzka, A. (2021). Computers in Human Behavior Prosociality and endorsement of liberty : Communal and individual predictors of attitudes towards surveillance technologies. *Computers in Human Behavior, 125*, 1–12. https://doi.org/10.1016/j.chb.2021.106938

Wnuk, A., Oleksy, T., & Maison, D. (2020). The acceptance of Covid-19 tracking technologies: The role of perceived threat, lack of control, and ideological beliefs. *PLoS ONE, 15*(9), 1–16. https://doi.org/10.1371/journal.pone.0238973

Zuboff, S. (2015). Big other: Surveillance capitalism and the prospects of an information civilization. *Journal of Information Technology, 30*(1), 75–89. https://doi.org/10.1057/jit.2015.5

Summary

The expansion of ICTs has tremendously enhanced life opportunities and made processes more efficient, and often safer. However, the advent of Big Data has nurtured the process of datafication of society, or the constant transformation of social processes into data. Critics illustrate how the datafication process not only creates new risks and uncertainties, but also enables new ways in which old uncertainties are reproduced.

In the thesis I apply Beck's Risk Society perspective to the study of datafication, with a twofold aim: on the one hand, the Risk Society theory can explain the uneven acknowledgment of risks unfolding within the datafied society; on the other hand, the process of datafication constitutes an interesting case to test the empirical grounding of the reflexive modernization thesis, according to which, as modernization progresses, technological progress is increasingly questioned. The first two empirical chapters deal with the process of risk definition and organized irresponsibility, from the perspective of individuals. In particular, I investigated trust dynamics and risk acknowledgment amidst datafication processes. The last two empirical chapters address mechanisms of risk stratification based on knowledge, by investigating educational gaps in the acknowledgment of datafication-induced risks and their conditionality on the spread of ICTs at the contextual level.

In Chapter 2 I focus on the impact of the controversy on online data privacy ignited in 2018 by the Cambridge Analytica scandal and by the introduction of the EU General Data Protection Regulation (GDPR) on trust in social media, while also exploring the endogenous vs. exogenous nature of this relatively new type of institutional trust. If trust in social media is endogenous and influenced by a critical evaluation of its functioning, the scandal would lower trust via framing. However, if trust in social media has its roots in exogenous factors, it would mainly flow from cultural determinants. This is addressed empirically by relying on a panel study as part of the Dutch wave of the European Values Study 2017, questioning a representative sample about their trust in social media before and after the controversy over online data privacy, and by using ordinal and multinomial regression models. Analyses suggest that trust in social media is distinct from other types of institutional trust, and strongly affected by cultural explanations. Nevertheless, the data breach turmoil did not strongly erode trust, challenging the 'worked and won' dynamic of trust in the Risk Society.

Chapter 3 is based on the results of a vignette experiment specifically designed to test the impact of the risks of privacy violation on the acceptability of a COVID-19 Health Pass among Dutch citizens. I argue that the acceptability of the COVID-19 Health Pass depends on informational norms regulating the exchange of information enabled by the technology, and on institutional trust which protects against uncertainty. The vignette experiment has been administered to around 1,500 respondents in the Dutch LISS panel in May 2021, and ordinal regression models are used to analyze the data. Results show large support for the measure, fostered by institutional trust, and not eroded by the privacy-intrusive features of the Pass. Findings indicate a tendency to underestimate the risks stemming from such technologies, at least in the presence of more tangible threats such as a global pandemic.

Chapter 4 discusses whether and why education affects e-privacy management, and whether these educational gaps vary following a country's degree of digitalization. I empirically test two sets of mechanisms, one derived from the digital divide and diffusion of innovations theories, the other from the reflexive modernization theory. The study employs Eurobarometer data and multilevel linear regression model. Findings suggest that the years spent in education positively affect e-privacy management, and that this effect is largely mediated by digital skills and internet use, and to a lesser extent by a reflexive mindset. The educational gap in e-privacy management narrows in more digitalized countries.

Finally, in Chapter 5 I investigate whether, and why, individuals express different levels of acceptance of surveillance depending on their educational level, and whether this relationship varies with the level of digitalization and globalization growth of their country. Additionally, I ask whether the type of surveillance (online surveillance vs cameras in public areas) conditions these differences. I build on two theoretical frameworks, the cultural backlash and reflexive modernization. I use data from the latest wave of the European Values Study (EVS) and implement multilevel multivariate regression models. Findings indicate that the lower educated individuals are more prone to accept online surveillance, due to their stronger authoritarianism and weaker reflexive mindset; however, there is no educational gradient in acceptance of video surveillance in public areas. Additionally, the countries' levels of digitalization and globalization expansion do not condition the educational gradient.

In the conclusions, I elaborate on how some elements of the reflexive modernization theory do not pass the empirical test, since a country's level of digitilization does not deepen knowledge-based stratification mechanisms, and individuals do not adjust their trust in data institutions once the risks they generate become visible. Directions for future research, as well as general limitations, are pointed out. Nevertheless, I also note how the Risk Society perspective is beneficial to better understand datafication processes, as findings indicate the success of organized irresponsibility dynamics, as well as the important role of knowledge as a risk stratification mechanisms.

Acknowledgements

Quid magis est saxo durium, quid mollius unda?
Dura tamen molli saxa cavantur aqua
(Ovidio, Ars Amatoria I:475–76)

Throughout the five and a half years that took me to build this work, I have encountered many hard stones. Beyond the many conceptual and analytical obstacles that writing a doctoral dissertation naturally entails, and beyond external circumstances (to name one, a global pandemic), this PhD forced me to face my personal weaknesses and vulnerabilities. Reaching the end of this journey, I think I can say that I was able to carve most of these hard stones and overcome the hurdles, thanks to some perseverance on my end, but mostly thanks to the steady support of many people around me. I am thankful beyond measure for your encouragement, patience, criticisms, and surely also the pushes that I, at times, needed to keep going.

First of all I want to thank my supervisor, Peter Achterberg, and my co-supervisor, Ruud Luijkx, for their guidance throughout the years. You challenged me to keep sharpening my thinking and writing, and I am very grateful for this. I also wish to thank the members of the PhD committee, Professor Veltri, Professor Wyatt, Professor Park, Professor van Reisen, and Professor de Koster, for taking the time to read through the thesis and provide insightful comments.

Next to the 'traditional' PhD trajectory, my academic experience has been shaped by the European Values Study. It has been an exceptionally valuable experience to learn how data comes to be in an international survey program setting. A huge thank you goes to my EVS colleagues throughout the years, with a special mention to the *avamposto italiano* for being great team members, to Vera Lomazzi and Tobias Gummer for welcoming me at GESIS, and to Evelyn Brislinger for teaching me the value of good data management.

Being surrounded by supportive colleagues who were always open to discuss ideas and answer questions has been essential. I hence wish to thank the members of the Department of Sociology at Tilburg University, and in particular Tim Reeskens and Christof Van Mol for giving me the opportunity to work with them, and my fellow PhD students for making many of my days easier to bear. I am also grateful to my ODISSEI colleagues, for welcoming me while I was still writing the last bits of the dissertation and for making me feel that my experience is valuable.

There have been moments in which I thought I had no energy left for carving, but luckily people around me did not let me give up. I am deeply thankful to Doctor Gianmaria Bartiromo for his professional help, and to the TechnoSoc writing group for helping me to structure my weeks when I needed it the most. I am also deeply grateful to my friends, some of whom have accompanied me since forever, some of whom I met during the journey that took me from Milano and Trento to Tilburg and Rotterdam. Thank you for always lending a ear, taking my hand, and giving me the occasional push.

I am also deeply thankful to my family for their unconditional love and support. Beatrice, being your big sister inspires me to always strive to be a better person. Mamma, Papá, not only you gave me a lot of opportunities that eventually led me to Tilburg and to the PhD, but you also taught me to be mindful of my online privacy, and therefore actively contributed to shaping this work. Thank you, for everything.

A special thank you goes to my paranymphs, for being on my side in all the possible ways. Thank you Elisa, it was quite a ride from our dear 94 to the

Aula of Tilburg University. And thank you Francesca, you and Sisma made me feel at home in Tilburg for the first time.

Last but certainly not least, I am grateful to Giovanni, my best friend, my partner, my husband. You have been there at all stages, always open to discuss ideas, cheering me up on the bad days, teaching me to celebrate victories on the good days. Thank you for believing in me, and making me see the soft but steady wave that carves the stones.

Tilburg, October 2022

This book has been produced in line with the EU GPSR guidelines about the safety of products.

The General Product Safety Regulation is the European Union's updated framework for ensuring that all consumer products, including books, are safe for consumers.

This book has been printed by Libri Plureos GmbH. The printer has issued safety certificates for the materials - like ink, paper and glue - being used.

The product identifier is: 9789403676302

The author is responsible for the content of the book, is the publisher of the works and bears full responsibility for it.

The book has been produced via Open Press Tilburg University. Open Press Tilburg University enables any author to share their stories with the rest of the world via printed books and ebooks and a broad distribution network.

Open Press Tilburg University will act as an intermediary in regard to questions about safety and will address them to the printer / author. Should there be any question in regard to the safety of the product, please contact us.

Open Press Tilburg University
Delftsestraat 33
3013AE Rotterdam
The Netherlands
info@mybestseller.com